Climate Change in Eurasian Arctic Shelf Seas

Centennial Ice Cover Observations

Ivan E. Frolov, Zalman M. Gudkovich,
Valery P. Karklin, Evgeny G. Kovalev and
Vasily M. Smolyanitsky

Climate Change in Eurasian Arctic Shelf Seas

Centennial Ice Cover Observations

 Springer

Published in association with
Praxis Publishing
Chichester, UK

Professor Ivan E. Frolov
Professor Zalman M. Gudkovich
Dr Valery P. Karklin
Dr Evgeny G. Kovalev
Dr Vasily M. Smolyanitsky
Arctic and Antarctic Research Institute (AARI)
St Petersburg
Russia

SPRINGER–PRAXIS BOOKS IN GEOPHYSICAL SCIENCES
SUBJECT *ADVISORY EDITOR*: Philippe Blondel, C.Geol., F.G.S., Ph.D., M.Sc., Senior Scientist, Department of Physics, University of Bath, Bath, UK

ISBN 978-3-540-85874-4 Springer Berlin Heidelberg New York

Springer is part of Springer-Science + Business Media (springer.com)

Library of Congress Control Number: 2009924773

Scientific editor: Andrey Proshutinsky
Translation editor: Professor Donald Rapp
English translator: Irina Solovieva
General editor: Vicky Cullen

Cover design: Jim Wilkie
Typesetting: OPS Ltd, Gt Yarmouth, Norfolk, UK

Printed in Germany on acid-free paper

Contents

Preface

The Arctic Seas ice cover has been regularly monitored since active development of the Northern Sea Route began in the 1930s. The monitoring is accomplished by a network of polar stations, using airplanes, ships, and, since the late 1960s, satellites.

Studies of long-term changes in the ice cover aim to address both economic and purely scientific objectives. Ice cover affects construction of ports and other onshore structures and operation of offshore platforms for production of hydrocarbons as well as their transportation to the mainland. In the science realm, it is impossible to understand the main mechanisms of Earth's climate changes—and hence predict future changes—without investigating long-term changes in the state of the Arctic Seas ice cover.

For the last few years, in connection with current climate warming, studies by Russian and other scientists have predicted a significant decrease in sea-ice extent in the Arctic and even its complete disappearance in the summertime by the end of the twenty-first century. This monograph presents results of studies of climatic system changes in the Arctic focused on the ice cover that do not justify such extreme conclusions.

Alternating periods of warming and cooling were typical in the Arctic during the twentieth century. The authors show that the duration of the main climate-shaping cycle of these fluctuations was about 60 years, but there were also 20- and 10-year cycles. The authors analyze the spatial–temporal peculiarities of these cycles and their influence on sea-ice extent variability.

They show relationships between long-term changes in area of the ice cover and climatic fluctuations in air temperature, atmospheric circulation indices, characteristics of water masses, and river runoff volume. There is also an analysis of possible natural causes of intra-secular climate fluctuations that influence the state of the Arctic ice cover (its area, thickness, concentration and multiyear ice edge). Based on the data presented for the twentieth century, the authors project Arctic Seas ice cover conditions for the twenty-first century: they expect that an oscillatory (rather

than a unidirectional) background of ice area changes in the Arctic Seas will be preserved during the current century, with a gradual increase by the 2030s and a subsequent decrease by the 2060s.

Many studies and international projects, such as the Arctic Climate Impact Assessment (ACIA), attribute the air temperature increase during the last quarter of the twentieth century exclusively to accumulation of greenhouse gases in the atmosphere. However, these studies typically do not account for natural hydro-meteorological fluctuations whose effects on multiyear variability, as this monograph shows, can far exceed the anthropogenic impact on climate.

Academician V. M. Kotlyakov
Director of the Institute of Geography
Russian Academy of Science

Figures

Tables

Abbreviations and acronyms

AARI	Arctic and Antarctic Research Institute
ALPI	Aleutian Low of atmospheric pressure
AR	Anticyclonic
AOBP	Arctic Ocean Buoy Program
AOO	Arctic Ocean Oscillation
AO	Arctic Oscillation
COWL	Cold Ocean Warm Land
CRU	Climate Research Unit (University of East Anglia)
CR	Cyclonic
ECO	East Canadian Oscillation
EOF	Empirical Orthogonal Function
GDSIDB	Global Digital Sea Ice Data Bank
GSAT	Global Surface Air Temperature
ICEX	ICe EXperiment
IICWG	International Ice Charting Working Group
JCOMM	Joint WMO/Intergovernmental Oceanographic Commission for Oceanography and Marine Meteorology
MIZEX	Marginal Ice Zone EXperiment
MY	Multi-Year ice
NIPCC	Nongovernmental International Panel on Climate Change
NAO	North Atlantic Oscillation
NEO	North European Oscillation
NP	North Pole
NEAZO	Norwegian Energy Active Zone
NSR	Northern Sea Route
PDO	Pacific Ocean Decadal Oscillation
SV	Singular Value
SLP	Sea Level Pressure

SA	Solar Activity
SSM/I	Special Sensor Microwave Imager
SAT	Surface Air Temperature
SLP	Surface Layer Pressure
TSI	Total Solar Irradiance
UV	Ultraviolet
WFC	World Fuel Consumption
WMO	World Meteorological Organization

Introduction

Arctic sea ice is an important climate parameter that regulates processes of heat, salt, and momentum exchange between the Arctic Ocean and the Arctic atmosphere. As part of the global climate system, Arctic sea ice directly influences the climate of the Northern Hemisphere and is also influenced by global climate changes. The major goals of this book are to describe the state and variability of the Arctic sea ice cover, to demonstrate methods for sea ice studies, and to describe and test hypotheses that will allow us to understand and predict future Arctic sea ice conditions. In order to reach these goals, we synthesize the data collected and experience gained by Arctic and Antarctic Research Institute (AARI) scientists during their more than 85 years of Arctic exploration.

Climate is usually defined as the "average weather" or as a statistical description that includes the mean and the variability of atmospheric, oceanic, and sea ice parameters over a period of time ranging from months to thousands or millions of years. The classical period for averaging is 30 years, as defined by the World Meteorological Organization (WMO). In meteorology, the most relevant parameters for characterizing climate are atmospheric variables such as temperature, pressure, precipitation, and wind. For sea ice, these parameters are ice concentration, thickness, drift, sea ice area, and sea ice extent.

Observational evidence indicates that significant climate changes have taken place throughout Earth's history. Geological data, isotopic evidence, dendrological and pollen analyses, vegetation and its fluctuations, glacier dynamics, lake-level variability, and instrumental observations over thousands or millions of years all confirm this.

Gribbin and Lamb (1978) describe four phases of climate change since Earth's last glaciation (approximately 10 kyr BP). The first postglacial climate warming (7–5 kyr BP) was characterized by a major decrease in glaciers and sea ice and a significant increase in mean air temperature, which was 2–3°C higher in summer compared to present conditions. The second phase was cooling that occurred in the

Iron Age (a period between 900 and 300 BC), distinguished by a decrease in air temperature, southward retreat of the northern forestry line, and changes in the precipitation regime. In the eleventh and twelfth centuries (or in the eighth to fourteenth centuries according to Borisenkov, 1982), there was an epoch of "small" climatic warming, which was characterized by favorable navigation conditions in the North Atlantic (and Viking colonization of the Greenland coast and part of North America). The average air temperature in this epoch was approximately 1.5°C higher than in the Little Ice Age (eighteenth to early nineteenth centuries) and slightly above current temperatures. This phase, often referred to as the Medieval Warm Period, was replaced by the Little Ice Age, when the coasts of Greenland and Iceland were bounded by sea ice, and glaciers expanded in the Alps and other regions. The surface water temperature in the North Atlantic at that time was 2–3°C lower than in the 1920s–1930s, the period of the first Arctic warming in the twentieth century (Lamb and Johnson, 1964). Zakharov in *Formation and Dynamics of a Modern Climate of Arctic Regions* (2004) reconstructed sea ice conditions in the eastern Barents Sea during the Little Ice Age by analyzing observations of Russian navigators during their cruises to Novaya Zemlya in the eighteenth and nineteenth centuries. He concluded that typical August ice conditions during that epoch corresponded approximately to sea ice conditions observed today in late June.

It should be noted that the Arctic climate described above was unstable. During both the cold and warm phases, there were interspersed shorter colder and warmer periods. The intensity of changes in various climate characteristics and their effects varied strongly from region to region. Under these conditions, it is very difficult to determine the major frequencies and magnitudes of climatic variations because these parameters depend on a variety of characteristics as well as the quality of the data. In this study, we express climate variability in the form of quasi-fluctuations at different frequencies (i.e., climate variability has a polycyclic character).

Monin (1969) and Monin and Sonechkin (2005) provide a detailed system of temporal-scale classification for weather and climate, with emphasis on important and robust climate changes at:

— Pleistocene glacial periods (hundreds of thousands of years)
— Inter-secular periods (from hundreds to thousands of years)
— Intra-secular periods (decades)

In this classification interannual and shorter signals are not related to climate change, which are often used in contemporary climate analysis publications. Orvig's (1973) analysis of Arctic climate variability concludes that the WMO climate definition (30-year mean) is not applicable to the Arctic because polar climate fluctuations are very large. Dobrovolsky (2000, 2002) supports the idea of stochastic climate variability and also supports Hasselmann's (1967) temporal climate classification, which distinguishes only two main ranges of atmosphere–ice–ocean system change, namely, synoptic and climatic, where variability with a period longer than one month is climatic. In this study, we assume that climate variability is a variability with a period 10-year or greater.

Twentieth century climate change research repeatedly led to projections ranging from either the complete disappearance of Arctic sea ice or, on the contrary, increases in ice area and thickness. Most of these projections were based on linear extrapolations of prolonged climatic tendencies accepted by investigators as permanent. In spite of well known failures of linear extrapolation of climatic data, extrapolation was repeatedly applied during the second half of the twentieth century, up to the present. Thus, after the "Arctic warming epoch" in the 1920s–1940s, some concern about the consequences of continued global warming was expressed (Budyko, 1969). However, the average temperature in the northern hemisphere began to decrease beginning in the middle of the twentieth century. This gave rise to concern about the possible extended continuation of this process (Gribbin and Lamb, 1978). Based on analysis of changes in ice conditions and air temperatures in the Arctic from the end of the 1960s to the mid-1970s, Volkov and Zakharov (1977) predicted further cooling and increased ice cover area in the Arctic Seas up to the 1990s. It was supposed that climatic and ice conditions by that time would approximate those that were observed in the Arctic at the beginning of the twentieth century.

But, again, nature prepared a surprise: a new warming event began in the middle of the 1970s, and by the middle of the 1990s Arctic ice conditions were the mildest of the twentieth century. The scientific community again emphasized the implications of "global warming" and predicted its catastrophic consequences. In many studies and international projects (e.g., Arctic Climate Impact Assessment, 2005), increased air temperature recorded during the last quarter of the twentieth century is attributed exclusively to accumulation of greenhouse gases in the atmosphere. In the opinion of supporters of the catastrophic consequences of the "global warming" scenario, escalating air temperatures are expected throughout the twenty-first century. Based on mathematical models that incorporate this continuous air temperature increase, a decrease in ice area is predicted up to the middle of the twenty-first century (e.g., Vinnikov *et al.*, 1999; Johannessen *et al.*, 2004). Published predictions range from the complete disappearance of Arctic Ocean ice to the onset of a new glacial epoch within a restricted time frame. All of these studies ignore natural hydrometeorological fluctuations, which, as this monograph shows, contribute to multiyear variability and can exceed by many times the anthropogenic impact on climate. This monograph is devoted to investigating the manifestations of natural fluctuations of sea ice extent and of other characteristics of climate on varying scales.

Joint scientific programs undertaken by scientists of different countries will contribute to further study of these problems. The atmosphere, the ocean, and sea ice were among the major topics of study included in programs undertaken for the 2007–2008 International Polar Year and its legacy for the period after March 2009, headed by the International Council for Science and the World Meteorological Organization. In addition to thematic work, Russian plans include undertaking annual large scientific expeditions onboard R/V *Akademik Fedorov* to deploy and support North Pole drifting research stations (NP-35 during September 2007–July 2008, NP-36 since September 2008) and to establish other new research bases in the Arctic.

These studies will make it possible to obtain more detailed knowledge of the Arctic and the Antarctic and to develop observation systems, thus taking steps forward in investigating the changes occurring in the climatic system as well as in understanding their major causes.

1

Arctic sea ice as an element of the global climate system

1.1 PATTERNS OF INTERACTION AMONG ARCTIC PROCESSES IN THE GLOBAL CLIMATE SYSTEM

In addition to the atmosphere, the global climatic system encompasses the global ocean and its sea ice cover as well as features on land that include glaciers, permafrost, rivers, and lakes. These system components continuously interact with each other. A number of studies identify patterns of such interaction, including *Formation and Dynamics of Modern Climate* of the Arctic regions (Alekseev *et al.*, 2004), which provides recent qualitative patterns of polar-process interaction in the global climatic system. Its contributing authors stress that the "Arctic is quite a sensitive part of the global climatic system" (p. 4).

Influences on the polar climatic system include:

— Solar radiation, which is partly regulated by the ozone layer
— Transport of carbon dioxide and aerosols from other areas that influence the heat balance of the atmosphere and the underlying surface
— Heat and moisture fluxes from the atmosphere of low and temperate latitudes
— Horizontal heat and salt exchange with the global ocean
— River runoff and iceberg discharge
— Freezing and melting of sea ice
— Accumulation and melting of glaciers
— Permafrost processes
— Convection processes in polar-region waters, including deep and shelf convection

In order to understand Arctic ice cover, it is important to not only enumerate climate-shaping processes but also to estimate the role of their anomalies in climatic changes of different scale. Due to complex relationships among the processes governing climate, this problem is extremely complicated. Solutions to some of its complexities are considered in chapter 6 of this monograph. A brief review of approaches to the problem that are available in the literature is presented below.

Alekseev *et al.* (2004) consider the global impact on climate of the Arctic to be transferred primarily through atmospheric circulation, controlling the heat and moisture transfer to high latitudes and their fluctuations within the interannual variability range. In addition, fluctuations of large-scale atmospheric circulation influence the inflow of warm and saline water to the North European Basin and further to the Arctic Basin and are manifested in the changes in sea ice extent.

> "An inverse impact of the Arctic on global climate change is connected with fluctuations of sea ice and fresh water export from the Arctic Ocean to the North Atlantic, which influence the total sea ice area change and the development of deep convective water sinking in the sub-Arctic and Arctic regions of the global ocean" (p. 7).

Because the formation of intermediate water in the North Atlantic depends on it, Nikiforov (2006) considers the overflow of dense, deep, near-bottom water across the Faroe–Shetland strait sill (Wyville–Thomson Ridge) to be significant in climate fluctuations. This overflow phenomenon influences the intensity of the Gulf Stream, determining the quantity and characteristics of warm Atlantic water flowing to the Arctic Ocean. Exchanges between the atmosphere and the Arctic Ocean are responsible for climatic fluctuations in the hydrometeorological ocean system regime.

Sea ice plays a large role in these exchanges, as has been observed by many scientists conducting both theoretical studies and data analysis, and discussed further in sections 1.2 and 1.3. The connection between sea ice (its thickness, area, and other parameters) and climate has long been recognized (Zakharov, 1996). The relationships of the ice thickness to the air temperature and other factors were described in the nineteenth century by Stefan (1891), and then confirmed by many empirical studies (e.g., Zubov, 1944). However, the existence of these relationships does not prove a climate-shaping role for ice, but only points to its dependence on climate.

The climate-shaping role of sea ice is determined by the presence of feedbacks (positive and negative) between the ice cover and processes at work in the atmosphere and hydrosphere. Sea ice influences climate on a variety of scales, from local to global. A brief review of the role of feedback mechanisms that determine the influence of ice cover on the global climatic system and its peculiarities in the Arctic is given below.

1.2 MAJOR EFFECTS OF ARCTIC SEA ICE ON THE CLIMATIC SYSTEM

The most obvious influence of snow and ice cover on climate is its reflectivity (albedo). The albedo of the snow-ice surface is known to change over a wide range: from 0.98 for freshly fallen snow to 0.10–0.30 for deep puddles, heavily polluted ice, and open water leads among sea ice floes. Instrumental data (Budyko, 1969) shows that the average albedo of the Earth-atmosphere system with ice cover equals 0.62, while for the ice-free areas it is about 0.30. A relatively high albedo value strongly decreases solar radiation absorption by the snow-ice surface. According to Brooks (1952), it decreases air temperature in the Arctic by several tens of degrees Celsius. Fluctuations in the ice cover area also change the average albedo of the Earth–atmosphere system, which affects the state of the global climatic system.

In addition to the effect of albedo on Arctic air temperature, the heat insulating effect of the ice cover has a large influence. The ocean-to-atmosphere heat flux through ice, including the latent heat of ice formation, is mainly determined by the vertical temperature gradient between the water surface and the air. This flux decreases with increasing thickness of ice and snow. An ocean covered by several years' accumulation of ice releases only a small amount of heat to the atmosphere in winter. Cracks and fractures resulting from dynamic processes in the ice play a significant role in this release of heat (Buzuyev et al., 1999); although insignificant in area (a few percent of the ice cover), these features account for about 50% of the heat flux from the ocean to the atmosphere (Makshtas, 1984).

Budyko (1962, 1966, 1968, 1969) employed several schemes for estimating the influence of polar ice on Arctic thermal conditions. His calculations showed that under ice-free conditions the mean annual air temperature in the Central Arctic would have increased by approximately 15°C compared to current conditions. The highest air temperature increase would have occurred at the coldest time of the year, while in the summer months it would not have increased more than several degrees Celsius.

Thus, the Arctic ice cover significantly decreases the air temperature above it and contributes to the increased horizontal gradient of the air temperature between the low and high latitudes of Earth's Northern Hemisphere. The atmospheric heat influx to the Arctic, which should increase with increasing ice cover area, depends on this gradient (a negative feedback). The role of the meridional gradient of the air temperature in forming the general planetary air flow from east to west in temperate latitudes is equally important for understanding climate change.

Air masses are transformed as they pass over various surfaces, and ice distribution plays an especially important role in these transformations. The available theoretical studies provide a mathematical description of air temperature transformation in a simplified formulation (Doronin, 1959; Nikolayev, 1963). The empirical data presented by Nikolayeva and Shesterikov (1970) are quite accurately approximated by hyperbolas, with parameters given by Appel and Gudkovich (1992).

Heat exchange between the atmosphere and the ocean changes especially sharply at ice edges (Vize, 1944a). As Budyko (1969) shows, its decrease at the ice edge extends in a slightly weaker form to temperate and even tropical latitudes due to the air temperature transformation over the open ocean. According to his calculations, the mean planetary temperature decreases more than two degrees near the earth's surface. The temperature decrease in the zone from the equator to 60°N compared to ice-free conditions ranges from 1.5° to 2.7°C, and in higher latitudes it increases sharply to more than 12°C.

The influence of changes in air temperature, which depends on the position of the ice edge, has not only a global but also a regional and even a local character. This is indicated in Teitelbaum (1977, 1979), where the problem of the effect of sea ice extent in the Arctic Seas on air temperature is solved using statistical methods. It is convincingly shown that at the beginning of the ice-melt period (May–June), the air temperature controls further decay of the ice cover, because the albedo value depends on its anomaly (see also Gudkovich et al., 1972). However, with the appearance of open water, the air temperature, which depends on the ratio of ice-covered/ice-free water, gradually becomes predominantly a result of sea ice extent.

The intensity of cyclonic (anticyclonic) activity in the atmosphere depends on energy drawn from the horizontal gradients of heat fluxes across the underlying surface and the air temperature above it (Pogosyan, 1972; Nikolayev, 1981; Nikiforov, 2006). These conditions usually occur near the ice edge. As shown by Treshnikov et al. (1967) and Bulgakov (1975), the ice edges in the Antarctic and the Pacific Ocean at the end of winter are usually located near sharp changes in convection depth. Abramov and Frolov (1987) employed a numerical model to calculate the heat loss from the surface of the Barents Sea in the fall–winter period. They showed that the mesoscale variability of sea–air heat exchange during fall–winter depends on water stratification at the ice edge and influences the location of average trajectories of extra-tropical cyclones that cross the sea in a zonal direction. The data obtained by Popov (2002) indicate that even such mesoscale phenomena as flaw polynyas can significantly influence the transformation of a thermobaric field over the northern polar area.

The influence of the ice cover on atmospheric circulation is manifested in such phenomena as oscillations in the ocean–ice cover–atmosphere system (Gudkovich and Kovalev, 2002a) (see Chapter 4 for more details). According to Zakharov (1996, 1997), the relationship between the sea ice extent of the Arctic Seas and the strength of the Arctic High also results from ice-cover influence on atmospheric circulation (Vize, 1940). This influence also extends to the prevailing trajectories of cyclones, which move southward with an increase in the Arctic High and northward at its decrease.

An extensive zone of decreased ice thickness observed in the Arctic Seas in winter is dependent on summer melting and subsequent ice export to the Arctic Basin. The heat flux to the atmosphere across this ice is slightly greater than that from the cold continents to the south and from thick multiyear ice to the north. Using a scheme of the average distribution of the calculated ice zones with a different time of formation from that of Gudkovich et al. (1972) and Gudkovich and Doronin (2001), Nikiforov (2006) calculated the heat flux to the atmosphere through first-year and younger ice. The average value of this flow was $500 \cdot 10^3$ kJ/m^2 for a season or $63 \cdot 10^3$ kJ/m^2 for each winter month. "Therefore it is not surprising that the area of the Arctic Seas is a 'highway' for the Atlantic cyclones frequently penetrating the East-Siberian Sea" (Nikiforov, 2006, p. 98). These cyclones form the Atlantic–Arctic pressure depression, which contributes to heat advection to the Arctic. Its development depends on processes in the North European basin and the ice state in the Arctic Seas (positive feedback).

The influence of ice cover on the exchange of gases between the atmosphere and the ocean is less evident. It is known that the concentration of greenhouse gases in the atmosphere, on which the intensity of long-wave heat emissions from the Earth–atmosphere system to space depends, is regulated by the processes of gas exchange between the atmosphere and the ocean. Gas exchange between atmosphere and ocean in ice-covered ocean areas is very limited. Therefore, an increase or a decrease in the ice cover area and ice concentration should be reflected in the concentration of greenhouse gases in the atmosphere resulting in further climatic changes (Golubev *et al.*, 2004). However, taking into account that the solubility of a gas in water decreases with an increase in water temperature and that these changes are in the opposite phase to changes in sea ice extent, the influence of anomalies in the ice cover area and water temperature in the ice-free region act in opposite directions. Corresponding changes in the biosphere also play a role in these processes.

1.3   STABILITY OF THE SEA ICE COVER IN POLAR REGIONS

Budyko (1969, p. 25–24) suggests the "possible existence of two climatic regimes in high latitudes, connected with the presence and absence of polar ice". Both regimes "are unstable, so that the ice cover can appear and disappear as the result of small changes in climate-shaping factors and even in the absence of these changes as a result of self-oscillation processes in the atmosphere–ocean–polar ice system". This proposal was confirmed by Rakipova (1962) and in other authors' studies (e.g., Saltzman *et al.*, 1981). However, Doronin (1968) and Zakharov (1976, 1977, 1978, 1981, 1996, 1997) and Zakharov and Malinin (2000) called attention to the relationship of ice cover extent and the impact of the underlying fresh surface water layer, beneath which there is a halocline layer in which the water density rapidly increases with depth. On the one hand, this layer constrains the heat content of the water in summer, but on the other hand, it restricts heat flux to the surface from deeper layers where there are currents carrying heat from lower latitudes as a result of winter convection and vertical turbulent exchange. These processes contribute to increased ice thickness and area in winter, resulting in a decreased probability that the ice will melt the next summer. Zakharov (1976, 1977, 1978, 1981, 1996, 1997 and Zakharov and Malinin (2000) provide convincing arguments in favor of the important role of the halocline acting as a shielding layer restricting heat flux to the surface. The absence of this layer limits the ice cover extent much of the year (January–May). This is confirmed by numerous observations of a sharp decrease in the rate of sea-ice extent increase in the North European Basin in the middle of winter—long before the time when heat loss from the surface begins to decrease. All this led Zakharov to the fundamental conclusion that "the most significant cause of climatic changes in sea ice extent in the ocean is changes in the vertical water structure in the upper ocean layer, rather than changes in thermal conditions in the atmosphere" (Zakharov 1996, p.183). Cases are noted (Malmberg, 1969) when the appearance of ice near the northern shores of Iceland was preceded by a significant decrease in salinity and temperature in the upper water layer.

The surface Arctic water mass forms as a result of mixing of freshwater (excess of precipitation over evaporation and continental runoff) with oceanic water flowing from the Atlantic and Pacific Oceans. Zakharov (1996) considers fluctuations in freshened surface Arctic water to result from disturbance of the freshwater balance of the Arctic Ocean. The incoming component of this balance is continental runoff, inflow of decreased-salinity water through the Bering Strait and precipitation while the discharge component is composed of runoff of freshwater to the Atlantic Ocean and evaporation (Serreze *et al.*, 2006, Ivanov, 1976). Obviously, the freshwater balance should influence the volume and extent of surface Arctic low salinity water, but other factors contributing to this process are also important. These include the volume and salinity of water entering the Arctic Ocean, predominantly relatively saline Atlantic water. Studies of average salinity changes in the upper layer of the Kara Sea using a balance model (Appel and Gudkovich, 1984) showed the role of possible anomalies of Barents Sea water inflow to the Kara Sea in these changes to be comparable with the influence of annual river runoff anomalies. Continental discharge to the Kara Sea comprises more than 25% of the river runoff to the Arctic Ocean (Ivanov, 1980). Hence, to solve the problem of the origin of anomalies in surface Arctic water, it is necessary to consider not the freshwater balance anomalies, but rather the corresponding salt balance anomalies. A satisfactory relation between the continental runoff volume to the Arctic Ocean from the coast of Asia and North America (*World Water Balance*, 1974) and subsequent sea ice extent of the North European Basin given in Zakharov (1996) does not provide a convincing argument in favor of a decisive role for continental runoff, because, as Zakharov points out, the data he uses were derived from calculations done by indirect methods. The anomalies of iceberg discharge were not taken into account in the calculations, and the series compared are short (25 years). The continental runoff in this calculation comprises only 42% of incoming freshwater. In addition, its influence on Arctic water extent is strongly complicated by changes in the Beaufort anticyclonic gyre system (Volkov and Gudkovich, 1967; Alekseev *et al.*, 2000; Nikiforov, 2006). A similar correlation of sea-ice extent with observed data on runoff from the largest rivers to the Arctic Basin seas, which supplies freshened Arctic water to the North European Basin, does not reliably confirm the relationship of sea-ice extent to continental runoff, as noted in Zakharov (1981).

It is important, however, to stress that justification of the role of the halocline in forming the sea ice cover of the Arctic Ocean excludes the possibility that a small increase in the incoming part of the Arctic heat balance can lead to relatively rapid disappearance of Arctic ice.

2

Long-term changes in Arctic Seas ice extent during the twentieth century

2.1 CHARACTERISTICS OF SEA ICE DATA IN THE ARCTIC SEAS

In general, the geographical terminology used in this book follows the Russian definitions published in Anon. (O). Treshnikov *et al.* (1967) define the Arctic Basin as a "near-pole abyssal basin, restricted by the continental slope." The Beaufort and Lincoln Seas are the marginal zones of the Arctic Basin. The North European Basin encompasses the Greenland, Norwegian, Barents, and White Seas as well as the Arctic Seas of Siberia (the Kara, Laptev, East Siberian, and Chukchi Seas). Baffin Bay, Davis and Smith Straits, Hudson Bay, and the straits of the Canadian Arctic archipelago compose the East Canadian region of the Arctic Ocean (e.g., Zakharov, 1996; Smirnov, 1974). Based on the major characteristics of the area's ice regime, the Greenland, Iceland, Norwegian, Barents, and Kara Seas are collectively known as the Nordic Seas, as proposed by Vinje (1998).

This study focuses on climatic changes in the region along the Northern Sea Route: the North European Basin, the Arctic Seas of Siberia, and the adjoining areas of the Arctic Basin (see Figure 2.1). The total area of this region is about 5 million km^2; it includes approximately 75% of the Arctic Ocean (Alekseev *et al.*, 2004). The sea ice edge very rarely extends beyond the marginal seas; thus, it can be assumed that the variability of the ice extent within this region is determined by the variability of ice extent in the North European Basin and the Arctic Seas of Siberia. Zakharov (1997) estimates that the North European Basin contributes 53% of Arctic Ocean ice extent variability for June–October, and the Arctic Seas of Siberia contribute 47%. Corresponding contributions in August are 26% and 74%, respectively. In winter (November–May), the main contribution to ice extent variability is almost equally shared by the Greenland and Barents Seas (48% and 52%, respectively).

Investigation of long-term changes in ice extent influenced by climatic variability requires observation series covering at least a century. This study draws mainly on data from observations in the Arctic Ocean east of Greenland, that is, from the

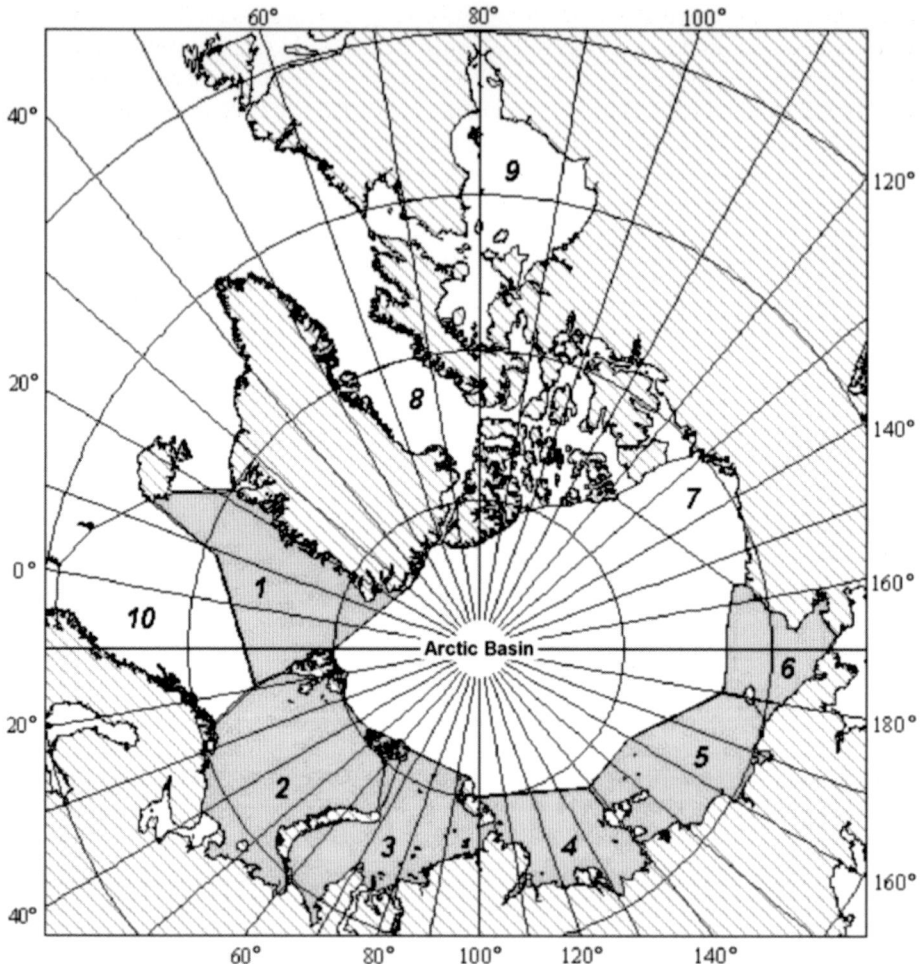

Figure 2.1. Boundaries of the Arctic Ocean and its seas: the Greenland (1), Barents (2), Kara (3), Laptev (4), East Siberian (5), Chukchi (6), and Beaufort Seas (7), Baffin Bay (8), Hudson Bay (9), and the Norwegian Sea (10).

Greenland Sea in the west to the Chukchi Sea in the east. Systematic data on the Greenland and Barents Seas cover the last third of the nineteenth century and almost the entire twentieth century (excluding the period of the Second World War). Regular observations in the Siberian shelf seas (Kara to Chukchi Seas) began in 1938 (Vize, 1940). The most reliable data on ice extent in the Arctic Seas are AARI airborne and satellite observations covering the period since 1940. The data on ice extent in the Arctic Seas in August for the period 1924–1939 were collected by Vize (1944a), who compiled all shipborne and airborne observations available at that time on ice edge positions during the initial period of active development of the Northern Sea Route. Several experimental "through" voyages from the Barents Sea to the Bering Sea along the Northern Sea Route (*Sibiryakov* in 1932, *Cheluskin* in 1933, *Litke* in 1934) were carried out. In 1935, the first through voyages of four cargo motor ships were made from west to east and from east to west. Episodic airborne ice reconnaissance in the Arctic began in 1924. Airplanes regularly carried out ice reconnaissance over the Kara Sea beginning in 1929, and over the Laptev Sea beginning in 1935. In 1938, this pioneering period of airborne ice reconnaissance ended. Since then, airborne ice observations have been regularly carried out along the entire Northern Sea Route (Vize, 1948).

Based on 1924–1939 sea ice and other data, Vize (1944a) investigated the correlation between the variability of the total ice extent of the Arctic Seas and indicators of the intensity of atmospheric circulation, such as atmospheric pressure and the total area influenced by the Arctic High. Vize's results were corroborated by later studies of Gudkovich *et al.* (1972), whose more extensive data covered a much longer time period. Findings from these investigations were used for development of methods for long-term forecasts of sea ice conditions along the Northern Sea Route (Gudkovich *et al.*, 1972), indirectly confirming sufficient reliability of Vize's data for use in the study of long-term (climate) variability in the ice extent of the Arctic Seas. From 1940 to 1979, ice charts were constructed from data collected by regular visual airborne sea ice reconnaissance; the charts for the period 1980–1992 were based on airborne and satellite observation data, and beginning in 1993, only satellite observations have been used in making the charts (Borodachev and Shilnikov, 2002).

Practically no data are available on Siberian shelf seas ice extent for the beginning of the twentieth century (1900–1923). In order to obtain a full 100-year data series on the ice cover of the Arctic Seas, an attempt was made to reconstruct them from a variety of sources containing descriptions of Arctic voyages at the beginning of the twentieth century. At the end of the nineteenth century and the beginning of the twentieth century, commercial, trade, and expedition vessels sailed in the Kara Sea, the eastern East Siberian Sea, and the Chukchi Sea. At the same time, the western East Siberian Sea and the Laptev Sea were visited more rarely. Observations from these voyages provided a basis for characterization of ice conditions in the studies of some Arctic investigators. Lesgaft (1913) compiled shipborne observations of Kara Sea ice conditions through 1911, and Nansen (1915) recorded annual descriptions of ice navigation conditions in the same sea from 1870 to 1913.

Substantial information on ice conditions in the Kara Sea and the eastern Arctic from Kolyma to the Bering Strait is contained in Itin (1933) and Sibirtsev and Itin (1936) for each year from 1900 to 1934. These authors analyzed multiyear variability of ice navigation conditions using a 5-point scale from ice-free years (1 point) to very severe years (5 points). Detailed data on the ice situation in the Arctic Seas during the period of the Arctic expeditions at the beginning of the century is presented in Vize (1948). Descriptions of ice conditions for separate years are contained in volumes on *Sailing Directions for the Kara, Laptev, East Siberian, and Chukchi Seas* published during the period 1935–1939 (Anon. (I, J, K, L)).

Charts showing ice conditions with the routes of ships and ice-edge positions are most valuable in these publications. In most cases, along with a description of the character of ice conditions (favorable, unfavorable), these studies include coordinates or orientation marks (islands, capes, and distances to them) and ice-edge positions along the ship routes. These mostly fragmentary data were plotted on the charts and then supplemented by expert assessment of the ice-edge position within the sea. Expert assessment was based on the characteristics of ice-edge positions under the main types of ice conditions—heavy, average, or light, determined on the basis of multiyear studies of the Arctic Seas ice regime using data for the period covered by reliable observations.

For 24 years at the beginning of the last century, the ice edge positions in August were reconstructed (and then the ice extents were calculated) in the Kara Sea for 10 years, in the Laptev Sea for 7 years, in the East Siberian Sea for 6 years, and, most importantly, in the Chukchi Sea for 18 years. For all other cases, the ice extent was calculated using mean monthly atmospheric pressure at regular grid points by a physical-statistical model developed on the basis of a simplified discriminant analysis. The model satisfactorily performed ice forecasts and calculations (Kovalev and Yulin, 1998). The correlation coefficient of the calculated data to the data reconstructed from published sources is 0.72. Thus, the combination of reconstructed data and observational materials made it possible to perform a comparative analysis of the changes in ice extent over the seas of a vast region (about $5 \cdot 10^6 \, \mathrm{km}^2$) throughout the entire twentieth century.

The seas under consideration can be divided into two significantly differing groups. The first group, including the Greenland and the Barents Seas, is characterized by the fact that part of this area remains ice free even in winter (Anon. (F)). The Norwegian Sea, which is also included in this group, is not considered here, since ice appears at its northern and western boundaries only in some years. Much of the White Sea is ice-covered in the second half of winter, and it is usually ice-free in summer; systematic data on the ice extent of this sea in February and March are available from 1951. For these months the average ice extent of the White Sea comprises only 10% of the ice extent of the Barents Sea. The correlation coefficient that characterizes the interannual changes in the ice extent of these seas from 1951 to 1994 is 0.50 (its significant value at $P = 95\%$ is not greater than 0.29). The changes in the ice extent of these seas are also quite similar: the ice extent in both seas slightly increases from the beginning of the 1950s to the end of the 1960s, after which it decreases until the middle of the 1990s. Therefore, we did not consider the state of the White Sea ice cover separately in this study.

The characteristics of the Arctic Seas ice regime for the first group indicate that the interannual variability of their ice extent is similar in all seasons. To determine seasonal differences, changes in the ice extent were examined in April–May when the ice extent is at a maximum and in August when it is close to the annual minimum.

The second group includes the Siberian shelf seas from Kara to Chukchi. Over much of the year they are mostly covered by very close ice (see sea ice term definitions under "Sea-Ice Nomenclature" in Anon. (M)), so the interannual variability of ice extent of these seas is observed only in the summer. For its characterization, the data for August were used, which closely correlate with the ice extent changes in July and September (Gudkovich et al., 1972).

2.2 SEASONAL AND REGIONAL CHARACTERISTICS OF ICE EXTENT TRENDS IN THE TWENTIETH CENTURY

In the past, scientists focused mainly on comparatively short-term changes (2–3, 5–7 years) in ice extent, while in recent years there has been increasing interest in investigating long-term variability (10 years and more), which reflect changes in Earth's climate. Detailed information about these changes is contained in the monograph by Zakharov (1996).

Plotting twentieth century ice extent changes in the Arctic Seas reveals a gradual decrease in ice extent from the beginning to the end of the century. These changes can be expressed by a linear trend, whose parameter (inclination of the straight line) is derived by a least-squares procedure. Using the value of this parameter and the series length, one can analytically derive the variance value (measure of ice extent scattering) described by a linear trend. The formula[1] for determining this variance is

$$\sigma_1^2 = \frac{a^2 n^2}{12},\qquad(2.1)$$

where σ_1^2 is the variance fraction, described by a linear trend; a is the trend parameter (km^2/year); and n is the series length (years).

Table 2.1 presents statistical data for each of the seas discussed above and their combinations, allowing comprehensive estimation of linear trends that characterize ice extent changes in the study region during the twentieth century.

Table 2.1 shows that the most significant twentieth-century changes in Arctic Ocean ice extent mainly occurred as decreases and increases at the boundary with the Atlantic Ocean—as was noticed earlier by Zakharov (1978, 1997). The largest changes described by the linear trends are evident in the ice extent of the seas of the first group (Greenland and Barents) in April (547 thousand km^2). Although similar changes in August are smaller (359 thousand km^2), the total changes during this period in the Greenland, Barents, and Kara Seas ("Nordic region") are comparable to the changes in April in the first two seas. The linear trend in the Laptev, East Siberian, and Chukchi Seas is an order of magnitude smaller. Thus, the most significant linear trend is in the ice extent of the Nordic seas. Its contribution both in April and August is typically greater than 30%. The share of the linear trend to ice extent variability in the seas located to the east of Severnaya Zemlya is only between 0 and 8%.

Figures 2.2 and 2.3 plot total ice extent changes in August for three "western" seas (Greenland, Barents, and Kara) and three "eastern" seas (Laptev, East Siberian, and Chukchi) in the twentieth century and their corresponding linear trends. The linear trend of ice extent for the western seas is much greater (4.3 times) compared to the eastern seas. One characteristic of the western seas is important: the decrease in

---

[1] Equation 2.1 is obtained by a simple integration for the ice extent L variance $\sigma_1^2 = (1/n)\int_1^n (L - \bar{L})^2\, dt$ by assuming that initial time moment t is located at the center of the series so that if $L = at + b$, we get $\bar{L} = b$, $(L - \bar{L}) \equiv at$ and $\sigma_1^2 \equiv (1/n)\int_{n/2}^{n/2}(at)^2\, dt$.

Table 2.1. Values of sea areas (S, thousand km^2); average ice extent (L, thousand km^2); ice exten changes for 100 years (1900–2000) (ΔL_{100}); variance ($\sigma_1^2 \times 10^6$ km^4), described by a linear trend; varianc of ice extent series ($\sigma^2 \times 10^6$ km^4); and their ratios (σ_1^2/σ^2)

Seas	S	L	ΔL_{100}	$\Delta L_{100}/L$	$\Delta L_{100}/S$	σ_1^2	σ^2	σ_1^2/σ^2
GS (IV)	1087	627	−303	0.48	0.28	10758	19304	0.56
BS (IV)	1388	857	−245	0.29	0.18	7645	23862	0.32
GS + BS (IV)	2475	1484	−547	0.37	0.22	24976	58714	0.43
GS (VIII)	1087	356	−95	0.27	0.09	759	6928	0.11
BS (VIII)	1388	201	−264	1.31	0.19	5802	17440	0.33
GS + BS (VIII)	2475	557	−359	0.64	0.15	10759	35517	0.30
KS (VIII)	830	444	−153	0.34	0.18	1959	23633	0.08
LS (VIII)	536	282	−38	0.13	0.07	122	9845	0.01
KS + LS (VIII)	1366	726	−192	0.26	0.14	3061	40184	0.08
ESS (VIII)	770	612	−37	0.06	0.05	113	11677	0.001
CS (VIII)	372	135	−45	0.33	0.12	171	2139	0.08
GS + BS + KS (VIII)	3305	1001	−505	0.50	0.15	21252	87166	0.24
LS + ESS + CS (VIII)	1678	1029	−120	0.12	0.07	1200	44158	0.03

Note: GS—Greenland Sea, BS—Barents Sea, KS—Kara Sea, LS—Laptev Sea, ESS—East Siberian Sea, CS– Chukchi Sea; IV, VIII—months of April and August

ice extent for the first half of the century was much more rapid than during the second half of the century. The formal calculation of parameters shows almost a sevenfold decrease in their value from the first half of the century to the second half.

Note that the linear trend parameter strongly depends on the time interval for which the trend is determined. However, as Gudkovich and Kovalev (2002b) show, in the presence of cyclic variability whose period is comparable to the time interval (see Sections 2.3–2.5), an assessment of the independent linear trend strongly depends not only on the series length but also on the choice of the starting point relative to the phase of cyclic variability. Ignoring this fact can lead to detection of a false trend or a strong distortion of the trend's value. The distortions are especially large in assess- ments of the trends for a time close to a cycle half-period. In such cases, they express a linear approximation of cyclic variability, which can be more strictly expressed by trigonometric functions (for example, a sinusoid). Even when we use sufficiently long time series, the linear trend can indicate both a unidirectional change in time and cyclic changes with periods exceeding the series length.

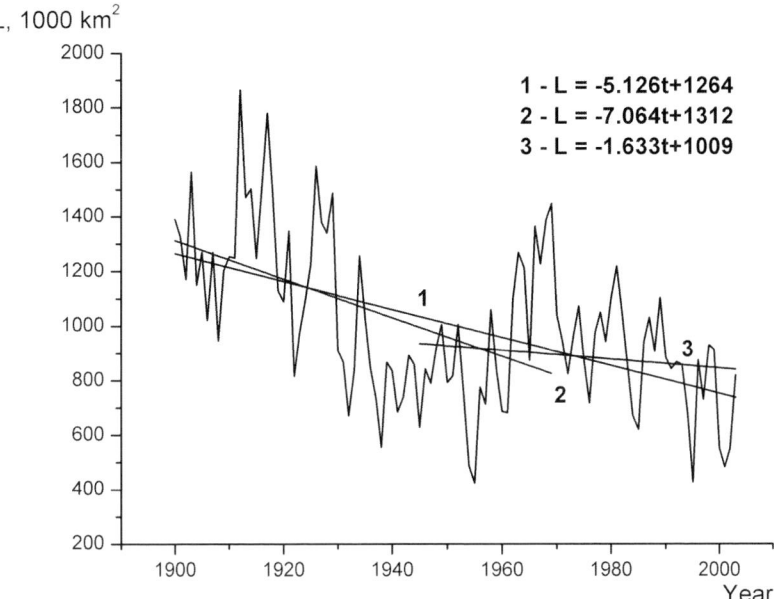

Figure 2.2. Linear trends in the changes in total ice extent in the Greenland, Barents, and Kara Seas for the period 1900–2003 (August). The inset shows the corresponding equations for linear trends: 1) 1900–2003, 2) 1900–1969, 3) 1945–2003.

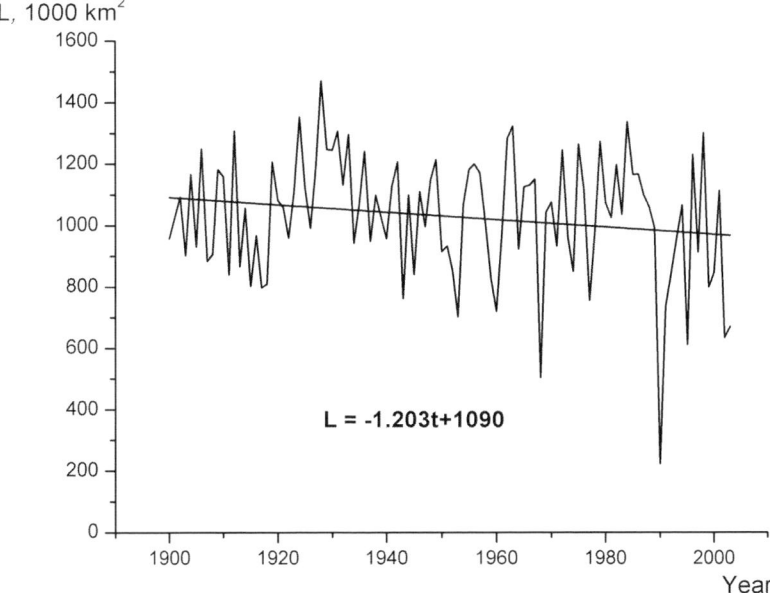

Figure 2.3. Linear trends in changes in total ice extent for the Laptev, East Siberian, and Chukchi Seas for the period 1900–2003 (August). The inset shows the equation for linear trends.

Errors in estimating the trend will be minimal if the trend parameter is calculated for a time interval between the neighboring maxima and minima of the most energy-intensive cyclic variability. For western region ice extent, 1900–1969 and 1945–2003 can be assumed to be such time intervals. Nevertheless, in this case the values of the corresponding linear trend parameters (-7.06 and -1.63) differ by more than four times.

Changes in the ice extent trend of the seas of the North European Basin in April for periods lasting for decades exhibited the same features; however, the differences in the trend values for the first and second halves of the century were much smaller. The trend of the ice extent in the eastern seas (Figure 2.3) is quite small and is not significant, as will be shown below. Its small increase in the second half of the century is determined exclusively by a large negative anomaly (-3.8σ), noted in 1990, and therefore it should be considered random in the study of climatic change.

A slower rate of ice extent decrease with time in the Nordic Seas was detected by Vinje (2000). By analyzing the ice extent changes in these seas for the period 1864–1998, he concluded that the total ice extent decrease in April in this region for 135 years, expressed by a nonlinear regression equation, was $0.79 \cdot 10^6 \, \text{km}^2$, or 33% of its initial value. Further, the rate of the ice extent decrease gradually deceased from $8 \cdot 10^3 \, \text{km}^2/\text{year}$ in 1880 to $3 \cdot 10^3 \, \text{km}^2/\text{year}$ in 1980. So, about 50% of the total sea ice extent decrease was observed during the four decades of the nineteenth century preceding the "period of Arctic warming."

What is the significance of these trends? We calculated the significance of the linear trends by means of standard procedures (Rozhkov, 2001; Stuart and Ord, 1994). Table 2.2 presents the estimates and confidence intervals at 95% significance level of linear ice extent trends in the Eurasian Arctic Seas. The calculations were conducted in general for the entire observation period (1900–2003), and also for two time intervals noted above for the western region seas.

Analysis of Table 2.2 shows that a negative trend in ice extent for August during the period 1900–2003 was observed for the Greenland, Barents, Kara, and Chukchi Seas as well as for the whole Eurasian Arctic. At the same time, a positive trend is not excluded but rather has a 95% probability for the Laptev and East Siberian Seas. Hence, the estimates of negative trends in ice extent for these seas are unreliable, similar to the total ice extent of the eastern region seas.

To determine the reliability of the linear trends in the western seas for the first and second halves of the twentieth century, the observation series were subdivided into two overlapping time intervals. The corresponding values are given in the lower rows of Table 2.2. These data indicate that the calculated linear trend coefficient for the first time interval is reliable, but it appears to be unreliable for the second time interval.

The seasonal and intra-secular changes in the linear trend of ice extent noted above suggest that this phenomenon should be studied in greater detail. The available observational data for the Barents Sea in the second half of the twentieth century allow us to calculate the values of the linear trend in ice extent of this sea for each month as listed in Table 2.3. As the table shows, the ice extent decrease in the Barents Sea in the second half of the twentieth century was observed only during the spring–

Table 2.2. Estimates of the linear trend coefficients ($y = ax + b$) of Eurasian Arctic Seas ice extent and their confidence intervals at 95% significance level

Sea (region)	Month	Linear trend coefficient		
		Estimate	Lower bound	Upper bound
Observation period 1900–2003				
Greenland	IV	−3.241	−4.007	−2.475
Barents	IV	−2.447	−2.905	−1.090
Greenland, Barents	IV	−6.482	−8.013	−4.951
Greenland	VIII	−0.954	−1.463	−0.446
Barents	VIII	−2.638	−3.324	−1.952
Kara	VIII	−1.533	−2.488	−0.578
Laptev	VIII	−0.381	−1.023	+0.261
East Siberian	VIII	−0.368	−1.068	+0.332
Chukchi	VIII	−0.453	−0.743	−0.163
Greenland, Barents, Kara	VIII	−5.126	−6.763	−3.488
Laptev, East Siberian and Chukchi	VIII	−1.202	−2.488	+0.084
Eurasian Arctic	VIII	−6.330	−8.367	−4.293
Observation period 1900–1969				
Greenland, Barents, Kara	VIII	−7.064	−10.393	−3.735
Observation period 1945–2003				
Greenland, Barents, Kara	VIII	−1.633	−5.154	+1.887

Table 2.3. Seasonal linear trend changes in ice extent for the Barents Sea for 1945–2000, thousand km^2/year

Months											
I	II	III	IV	V	VI	VII	VIII	IX	X	XI	XII
+3.50	+2.45	−0.41	−1.65	−1.37	−0.99	−0.80	−0.96	−1.01	+3.56	+3.11	+0.31

summer period (from March to September). During the autumn-winter period (October to February), when there is intense formation of young ice, a significant positive linear trend occurs in the variability of ice extent (Table 2.3).

Buzin (2006) found a similar trend for the Barents Sea and its northeastern sector for 1928–2003. It should be noted that the linear change in mean annual ice extent was close to zero.

Ponomarev *et al.* (2003, 2005) detected similar behavior in seasonal changes in twentieth-century climatic trends of surface air temperature in the middle and temperate latitudes of northeast Asia. Significant warming in winter was accompanied by noticeable cooling in summer. So, the seasonal changes in the trends of this parameter were opposite to those observed in the Barents Sea. The causes of these anomalies require further investigation.

2.3   THE POLYCYCLIC CHARACTER OF LONG-TERM CHANGES IN ICE EXTENT

The interannual changes in the ice extent of the Eurasian Seas of the Arctic Ocean appear to have a polycyclic character. The frequency structure of these changes revealed by several investigators is characterized by significant peaks for periods of 2–3, 5–7, 8–12, about 20, and 50–60 years (i.e., Volkov and Sleptsov-Shevlevich, 1970, 1971; Gudkovich *et al.*, 1972; Karklin, 1978; Karklin *et al.*, 2001; Karklin and Teitelbaum, 1987).

Although there are a variety of causes for the temporal structure of multiyear variability of hydrometeorological characteristics (including ice), the cyclic properties can be studied using methods of latent periodicities, such as periodogram and spectral analyses. Both methods yield more accurate results when analyzing series that present a sum of harmonic variability. During analysis of cyclic variability characterized by changes in duration (within some limits) and amplitude, the dominating periods (on the periodogram) or frequencies (on the spectrogram) can be "fuzzy" within the cycle's variability.

Our analysis of the spectrograms of ice extent variability in each of the seas under consideration revealed significant differences among them: a typical characteristic of the three western seas is significant low-frequency variability, while the three eastern seas exhibit relatively high-frequency variability. Therefore, to illustrate the temporal (frequency) structure of the variability of ice extent, analogous to the analysis of the linear trends, the Eurasian Arctic Seas were combined into two groups: western, including the Greenland, Barents, and Kara Seas, and eastern, encompassing the Laptev, East Siberian, and Chukchi Seas. The total ice extent in each of the groups was calculated on the basis of data for the period 1933–2003. The beginning of this period coincides with the start of regular airborne ice reconnaissance in the Arctic Seas; therefore, calculations for the latter two-thirds of the century offer higher reliability than calculations for the first third. The two available data series were subjected to spectral analysis with self-correlation functions calculated after 50 years, which is sufficient for distinguishing the frequencies in the low-frequency spectrum

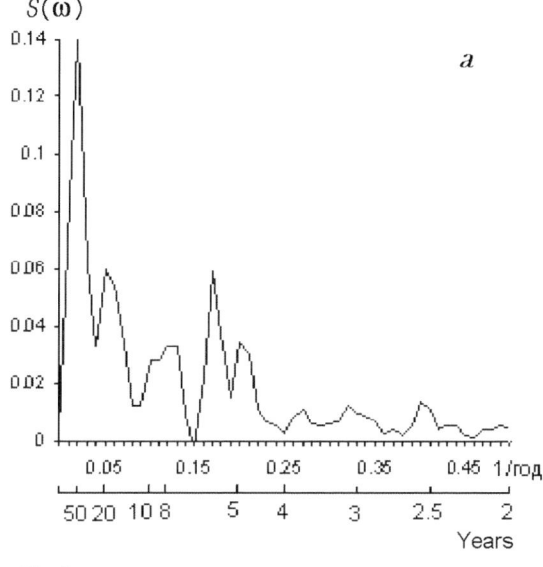

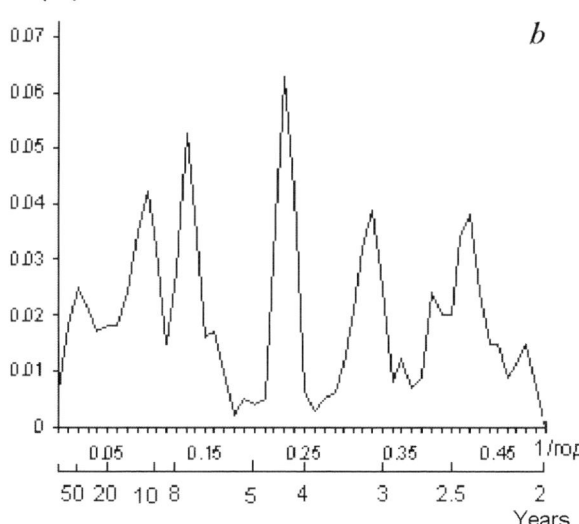

Figure 2.4. Functions of the spectral density of variability in total ice extent during August in the western (Greenland, Barents, and Kara) seas (*a*) and in the eastern (Laptev, East Siberian, and Chukchi) seas (*b*).

region. The spectra characterizing both groups are presented in Figure 2.4, where cycles lasting about 50–60 and 20 years play a significant role in forming the structure of multiyear variability of ice extent in the western seas. Their total contribution to the total variability exceeds 30% and is much greater than the contribution of the same cycles in the eastern seas (Table 2.4). At the same time, in the eastern seas, a more significant contribution to the total variability occurs in cycles of about 9–12 and 7–8 years; their contribution is twice as large as the contribution of the cycles of same duration in the western seas (Table 2.4).

Table 2.4. Percentage contribution of the main frequencies to the variability of total ice extents in August in the western (Greenland, Barents, and Kara) and eastern (Laptev, East Siberian, and Chukchi) seas.

Region	Month	Linear trend	Frequency, 1/years (cycles, years)				
			0.01–0.03 *(50–60)*	*0.04–0.06* *(20)*	*0.08–0.11* *(9–12)*	*0.12–0.15* *(7–8)*	*>0.17* *(2–5.5)*
Western	IV	43	5	4	4	–	–
Western	VIII	24	17.5	13	6.5	7	32
Eastern	VIII	3	7	5	12.5	13.5	59

Cycles lasting 2–5 years play the main role in ice extent variability in the eastern Arctic Seas, forming the interannual variability, whose contribution comprises almost 60% of the total variability. The role of these cycles in the western seas is less significant (Table 2.4).

Some components of long-period changes in ice extent appear to have a greater role in climatic changes than others (see Table 2.4). In determining the variance, if we exclude the high-frequency variability not related to climate changes, we obtain different estimates. Thus, five-year smoothing of the initial series of ice extent for the western region in August (periods with variability up to five years are excluded), the variability decreases by 62%. Hence, the total contribution of 50–60-year and 20-year cycles to the long-term variability of ice extent of the region increases to almost 50%. The total contribution of these cycles and the linear trend comprises more than 88%.

A wavelet-analysis method and software developed by Torrence and Compo (1998) was applied to estimate the temporal variability of the spectral structure of ice extent. Assuming that the period of wave fluctuation (forms given in Figure 2.5) is 15 years and using a sampling of ice extent values for the period 1950–1964, one can assess the correlation coefficient between this type of wave (and the period) and ice extent data. Moving the starting point forward yields consecutive estimates of the energy of the variability with a period of 15 years for the intervals 1951–1965, 1952–1966, etc.

Similarly, we can vary the elementary wave period and successively derive estimates of the energy of variability for scales of 2, 3, 4, 5 years, etc. Two types of elementary waves were used in the analysis: the "Morlet wave" (Figure 2.5a) to reveal typical spectral components of long-period variability and the Mexican hat (Figure 2.5b) to reveal the temporal structures of spectrum variability.

An important component of the analysis is checking the significance of the derived amplitudes of variability. The following method of assessing the significance is also proposed in Torrence and Compo (1998). Using the Monte Carlo method, red noise is generated with amplitude equal to series variance. By criterion χ^2, the red noise amplitude of 5% significance is determined. Then, if the amplitude of variability has a significance level greater than 5%, it can be said that at the given significance level the fluctuation of this period is probable.

We use the same approach to interpret results of the wavelet analysis as Monin and Sonechkin (2005). Figure 2.6 (see color section) presents wavelet-spectrum calculations of ice extent in August for 1900–2003 for six Eurasian Arctic seas (Greenland to Chukchi). For convenience of interpretation, the left-hand column (Figure 2.6a) presents the initial ice extent time series. The center column (Figure 2.6b) contains the results of wavelet-transformation of ice extent series in the form of the amplitude of variability (in 1000 km^2) with a sign, and as a basic component, a wave fluctuation of the "Mexican hat" type is used. The right-hand column (Figure 2.6c) contains the total spectrum of ice extent variability, which was also estimated by wavelet-transformation, where, as a basic component, a different wave fluctuation of the "Morlet wave" type is used.

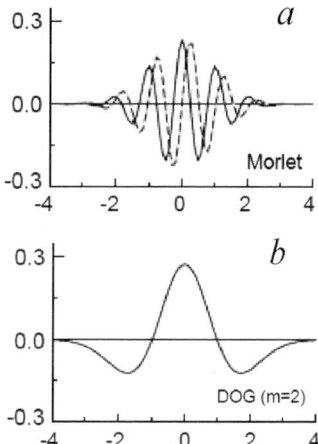

Figure 2.5. Forms of elementary wave variability used in the wavelet analysis. The vertical axes are dimensionless amplitudes of variability and the horizontal axes are dimensionless shifts in time.

An analysis of Figure 2.6b shows that the long-period spectrum structures of the seas of the western and eastern sectors of the Eurasian Arctic are different. For the Greenland, Barents, and Kara Seas, approximately synchronous 50–60-year variability is well distinguished with maxima near 1910 and 1970 and minima near 1940 and 2000. On the contrary, for the Laptev, East Siberian, and Chukchi Seas, shorter variability of 20–30 years and less is apparent. All the seas are also characterized by short-period variability of 8–10 years. The spectra of the western sector and the Eurasian Arctic seas are similar in general to the structure for the Greenland and Kara Seas while the spectrum of the eastern sector is similar to the structure for the Laptev and East Siberian Seas. Analysis of the total ice extent spectra (Figure 2.6c) yields similar results. It is interesting to note that ~100-year and ~60-year variability is statistically significant only for the Barents Sea and ~60-year variability for the Kara Sea. The latter are also statistically significant in the western sector and Eurasian Arctic spectra. According to Monin and Sonechkin (2005), ~100-year variability ought to be assumed as artificial.

2.4   THE "60-YEAR" CYCLE AND ITS ROLE IN ICE EXTENT CHANGES IN VARIOUS REGIONS

While the history of Arctic sea ice extent variability in the twentieth century is characterized by a negative linear trend, there were prolonged periods of time when sea ice extent showed more or less stable increases or decreases that resulted in positive or negative trends, respectively. Some studies (Zakharov, 1996; Mironov,

2004; Zubakin, 1987; Vinje, 2000) calculated alternating-sign linear trends for such intervals. These data allow us to coarsely assess the duration of the corresponding cycle of ice extent change over 50 to 60 years.

The presence of such variability both in the western and eastern regions allows us to distinguish three typical epochs: a decrease in ice extent in the first part of the twentieth century and its subsequent increase in the late 1960s–early 1970s, which was again replaced by a decrease in ice extent during the last three decades. The boundaries of the indicated time intervals and the intensity of the changes varied slightly from region to region and from winter to summer.

In Karklin *et al.* (2001), data on interannual changes in the total ice extent of the Siberian Arctic Seas in the twentieth century were approximated by a polynomial to the sixth power. The curve thus obtained well reflects the long-period variability of the ice extent in the region under consideration. The half-century wave period, identified by these authors for the first time, lasts 55–60 years, which is quite close to the rough estimates given above. Similar variability was detected in air temperature changes in a zone from 72°N to 87°N and in recurrence of the main atmospheric circulation forms identified by Vangengeim (1935) and Girs (1960). Karklin *et al.* (2001) noted that the long-period ice extent variability in the Arctic Seas, unlike shorter-period variability, does not indicate an opposition in phase between the western and eastern regions, which may testify to their common nature.

Figures 2.7–2.9 show a similar approximation of the changes in total ice extent in the Greenland and Barents Seas in April and in the total ice extent of three western

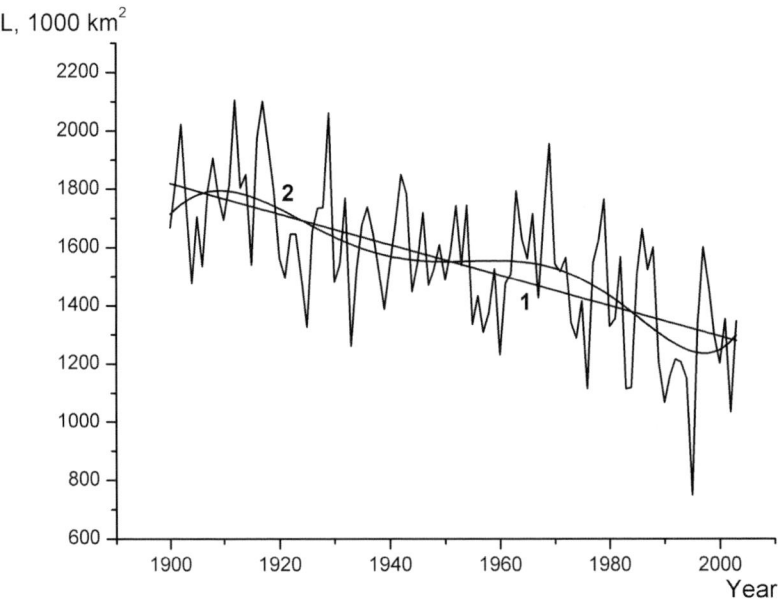

Figure 2.7. Variability of total ice extent in the Greenland and Barents Seas for the period 1900–2003 (April): 1) linear trend, and 2) polynomial trend.

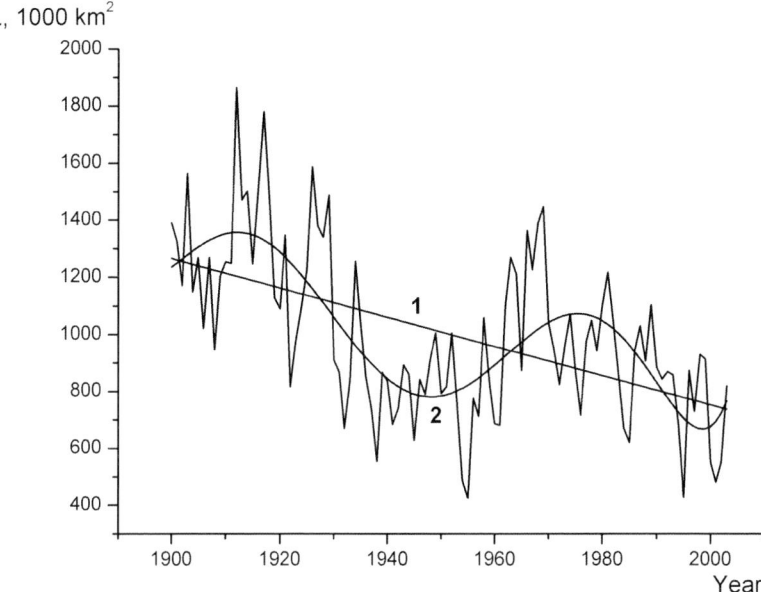

Figure 2.8. Variability of the total ice extent in the Greenland, Barents, and Kara Seas for the period 1900–2003 (August): 1) linear trend, and 2) polynomial trend.

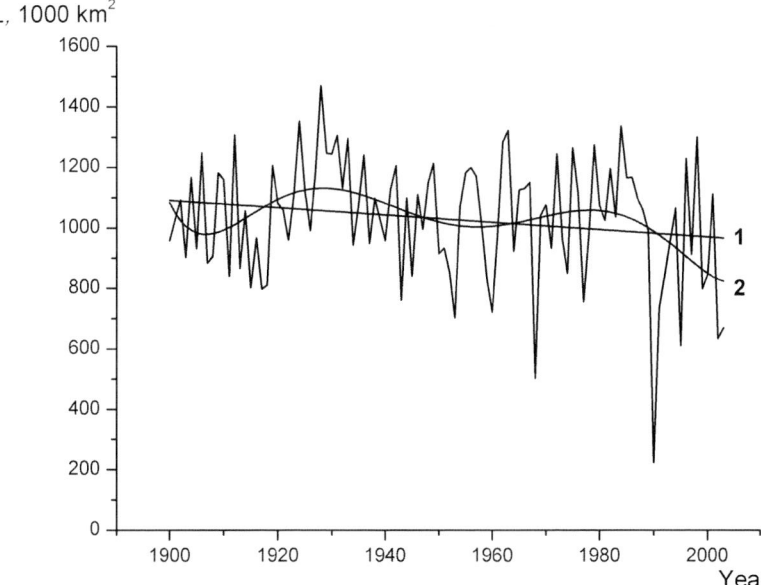

Figure 2.9. Variability of total ice extent in the Laptev, East Siberian, and Chukchi Seas for the period 1900–2003 (August): 1) linear trend, and 2) polynomial trend.

and three eastern seas in August. All three figures have common features: a negative linear trend from the beginning to the end of the twentieth century and long-period (55–60-year) variability. However, the amplitudes of the variability differ significantly: the phases of variability between the western and eastern seas are slightly displaced.

These figures show examples of data plots that can be used to estimate the "50-year" variability of ice extent in various seas and regions. The average amplitudes of variability under consideration, determined at the moments of the largest polynomial curve deviation from a linear trend, allow calculation of ice extent variance, as created by the wave in question, depicted by a curve, using the equation

$$\sigma_{60}^2 = A_{60}^2/2, \qquad\qquad (2.2)$$

where A_{50} equals the average amplitude of the 50–60-year wave.

This equation is based on changes in the harmonic curve, which can differ slightly from the real fluctuation. However, its advantage is that it allows an estimate of variance for the full period, whereas the real fluctuation can include the arbitrary parts of this period, which will influence the value of the variance. The results of calculations using Equation 2.2 for various seas, regions, and seasons are presented in Table 2.5. It shows the small effect (3–6%) of a "60-year" cycle in total ice extent variability of the seas in the North European Basin (Greenland, Barents) in winter and of the seas located to the east of Severnaya Zemlya in summer. This fluctuation is largest (up to 20%) in the Greenland Sea in summer.

The presence of such a clear cycle in ice extent changes in the Nordic Seas region influences estimates of the linear trend describing these changes. As Figure 2.7 shows,

Table 2.5. Average amplitudes of "60-year" components of ice extent and corresponding variance, along with its contribution to the total variability of the ice extent of the seas

Sea, month	A_{50}, thousand km^2	σ_{50}^2, 10^6 km^2	σ_{50}^2/σ^2, %
Greenland, April	35.3	623	3.2
Barents, April	55.9	1562	6.5
Greenland + Barents, April	77.9	3034	5.2
Greenland, August	41.2	849	12.3
Barents, August	81.4	3313	19.0
Greenland + Barents, August	111.7	6238	17.6
Greenland + Barents + Kara, August	175.0	15312	17.6
Kara, August	97.5	4753	20.1
Laptev, August	30.0	450	4.6
East Siberian, August	38.0	722	6.2
Chukchi, August	18.0	162	7.6

the phase of the "60-year" variability characterizes the conditions under which positive ice extent anomalies were noted at the beginning of the century and in the 1970s, and negative anomalies were noted in the 1940s and at the end of the century (two waves). Gudkovich and Kovalev (2002) show that these conditions lead to the appearance of a false negative linear trend. The calculations indicate that 10% of the variance described by the corresponding trend (Table 2.1) expresses the false trend influence.

2.5 20- AND 10-YEAR CYCLES AND THEIR ROLE IN ICE EXTENT CHANGES

In addition to a "60-year" cycle, shorter cycles lasting about 20 and 10 years appear to be present in long-term changes in ice extent. We studied these cycles using a periodogram analysis of ice extent time series in the seas under consideration for 1933–2003, i.e., for the period of the routine monitoring of ice conditions in the Eurasian Seas by air reconnaissance (up to 1992) and satellite information analysis (since the mid 1960s) (Borodachev and Shilnikov, 2002).

　　To exclude the influence of short-period variability, all series were first subjected to smoothing by a five-year running mean procedure. The results of this process are shown in Figure 2.10. The various seas and seasons depicted in the periodograms in this figure have common features: significant increases in the amplitude of variability within period ranges of 18–22 years and 8–13 years. However, the ratios between the amplitudes of first- and second-range variability are different for different seas.

　　Table 2.6 provides information on the amplitudes of these and other variations and their influence on ice extent interannual changes. The variance of each wave was determined by a formula similar to Equation 2.2. Account was also taken of the fact that the five-year running smoothing data filter decreases the values of the distinguished amplitudes (by 35% for a 10-year wave and by 10% for a 20-year wave). As Table 2.6 indicates, the amplitude of 20-year variability in the western seas (from the Greenland Sea to the Kara Sea) is much higher than that of 10-year variability (both in winter and summer). This ratio decreases in general from west to east. In the eastern seas (from the Laptev Sea to the Chukchi Sea), the differences in amplitude are much less, and in the Laptev and the Chukchi Seas the amplitudes of 10-year variability are larger than those of 20-year variability. The contribution of the former to the total ice extent variance for the western seas (23%) is notably less than for the eastern seas (38%), and vice versa for the "20-year" variation: in the western seas, it comprises 13%, on average, and for the eastern seas, about 7%.

2.6 SHORT-PERIOD VARIABILITY OF ARCTIC SEAS ICE EXTENT

Although the short-period (2–3-year and 6–7-year) cyclic variability of ice extent of the Siberian shelf seas is typically considered "noise" in comparison to long-period climatic variability, this obscures some significant changes in Arctic climate. The

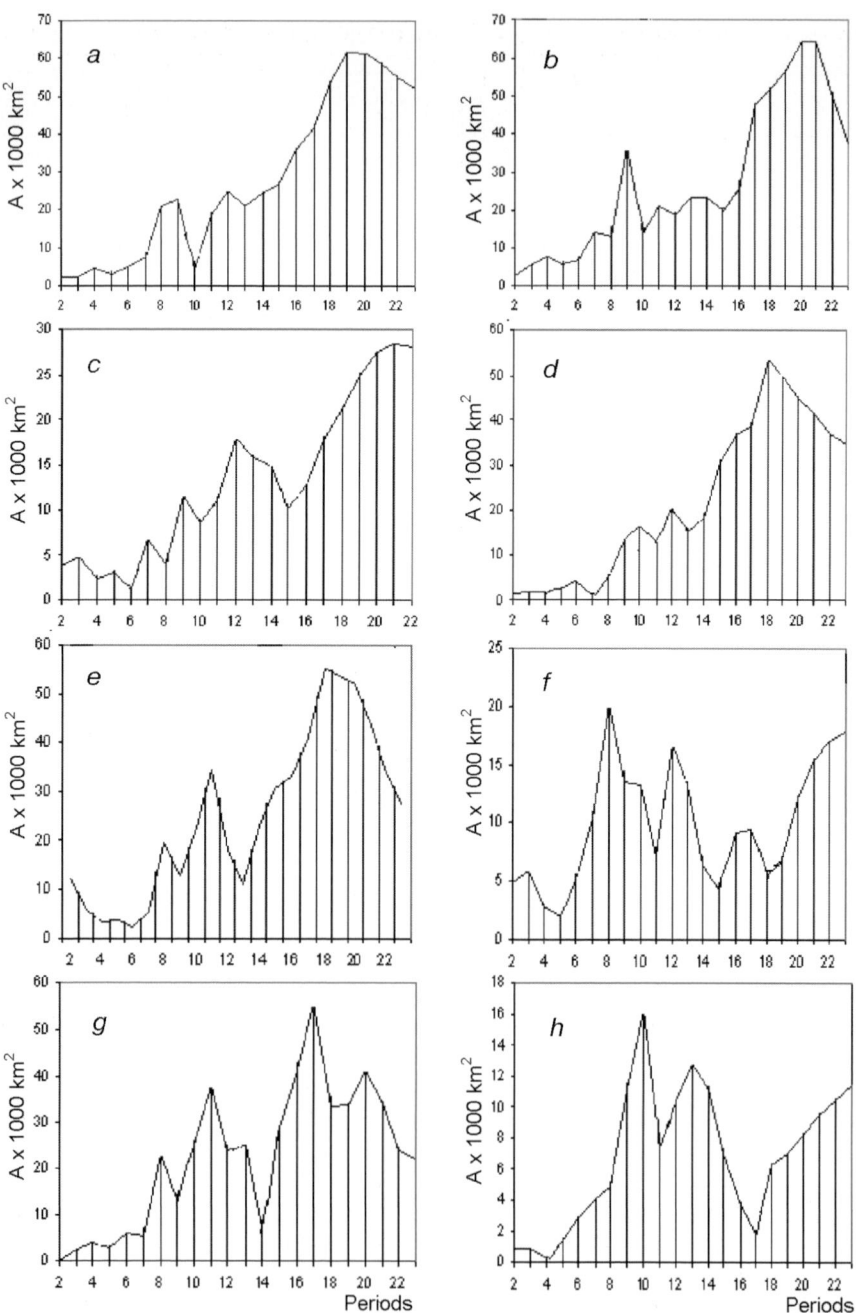

Figure 2.10. Periodograms of the variability of ice extent in the Arctic Seas, smoothed by a five-year running mean procedure for 1933–2003: *a*) and *b*) show the Greenland and Barents Seas, respectively, in April; for August, *c*) shows the Greenland Sea, *d*) the Barents Sea, *e*) the Kara Sea, *f*) the Laptev Sea, *g*) the East Siberian Sea, and *h*) the Chukchi Sea.

Table 2.6. Statistical characteristics of "20-year" and "10-year" components of ice extent variation

Seas	Month	A_{20} 10^3 km^2	A_{10}, 10^3 km^2	A_{20}/A_{10}	σ_{20}^2, 10^6 km^4	σ_{10}^2, 10^6 km^4	η_{20} (%)	η_{10} (%)
Greenland	IV	68.6	31.8	2.16	2353	506	12.2	2.6
Barents	IV	71.1	39.5	1.80	2528	780	10.6	3.3
Greenland	VIII	30.9	19.3	1.60	477	186	6.9	2.7
Barents	VIII	59.9	24.2	2.48	1794	293	10.3	1.7
Kara	VIII	60.9	37.0	1.65	1854	684	7.8	2.9
Laptev	VIII	19.2	26.0	0.74	184	338	1.9	3.4
East-Siberian	VIII	54.2	38.4	1.41	1469	737	12.6	6.3
Chukchi	VIII	12.4	18.8	0.66	77	177	3.6	8.3
Greenland + Barents + Kara	VIII	113.5	80.0	1.42	6435	3200	13.0	6.5
Laptev + East-Siberian + Chukchi	VIII	80.0	74.5	1.07	3200	2775	7.2	6.3

Note: η_{10} and η_{20} indicate contribution to the total "20-year" and "10-year" variability, respectively.

contribution of these cycles to the interannual variability of ice conditions in these seas and their areas (Gudkovich et al, 1972) is comparatively large; therefore, in the 1970s, a great deal of attention was devoted to their study in connection with development of ice forecasting methods. Studies by Volkov and Sleptsov-Shevlevich (1970, 1971) revealed the spatial-temporal structure of a "two-year" standing wave, whose loop was located in the Severozemelsky region, and its node in the vicinity of the New Siberian Islands. The variability of quasi-two-year anomalies in ice extent of the Siberian shelf seas was analyzed by Karklin (1977), and an assessment of the contribution to total ice extent variance, which changes from 20–50%, was carried out.

The publications of Maksimov (1960), Kovalev (1960), Volkov and Sleptsov-Shevlevich (1971), Gudkovich *et al.* (1970), and Karklin (1977, 1987) are devoted to investigation of a "7-year" cycle of ice extent changes. It was revealed that this variability is maximal in the eastern Laptev Sea (east of 125°E), where its range is 30–40% of the region's area. It was also observed that a "7-year" wave is gradually displaced from east to west traveling over 4–5 years from the Chukchi Sea to the Kara Sea, which partially explains the existence of opposite ice extent conditions between the seas of the western and eastern sectors of Eurasian Arctic.

The results obtained in these studies do not provide a complete understanding of the spatial and temporal effects of this cycle on the ice extent of the Arctic Seas. We analyzed this variability using the ice extent series of the Arctic Seas in August for the period 1933–1999. The filtration (weight) function proposed by Leith and Holloway (1958) was used to isolate 6–7-year variability (the cycle duration varies within 5–8 years; see Table 2.7), and to exclude 2–3-year variability as well as variability of 11

Table 2.7. Duration of short cycles and the number of cases of their occurrence during the period 1933–1999

Region	Cycle duration, years				
	4	5	6	7	8
Southwestern Kara Sea	–	2	4	1	–
Northeastern Kara Sea	1	2	3	1	–
Western Laptev Sea	–	1	4	2	–
Eastern Laptev Sea	–	3	1	3	–
Western East Siberian Sea	1	–	3	2	–
Eastern East Siberian Sea	1	2	–	2	1
Chukchi Sea	–	1	2	1	2

years or more from multiyear variability of ice extent. Table 2.7 shows that the cycle most often occurs for 5–6 years in the western regions of the Arctic Seas (from the Kara Sea to the western Laptev Sea), while in the eastern regions, the number of cycles lasting 6–7 years increases. In the total ice extent of the Arctic Seas, the 7–8-year cycles prevail.

The amplitude of the cycle is unstable and can change several times in each of the regions (Table 2.8). The maximum amplitudes are observed in the Laptev Sea. In the

Table 2.8. Amplitudes of 6–7-year cycles and their effects on multiyear variability of the Arctic Seas ice extent

Region	Amplitudes (thousand km^2)			Contribution to dispersion (%)
	Average	Maximum	Minimum	
Southwestern Kara Sea	35.5	50.9	17.4	29
Northeastern Kara Sea	32.7	56.9	9.9	14
Western Laptev Sea	27.4	57.3	8.7	26
Eastern Laptev Sea	42.5	80.4	10.0	41
Western East-Siberian Sea	41.0	69.7	11.3	31
Eastern East-Siberian Sea	25.6	45.2	10.2	18
Chukchi Sea	27.5	39.1	10.4	31
Total sea-ice extent	*145.5*	*243.3*	*65.2*	32

course of a 6–7-year cycle, the ice extent in this sea can change by more than 50%. To the west and east of the Laptev Sea, the cycle amplitudes decrease.

The contribution of 6–7-year variability is significant and ranges from 14 to 41% (Table 2.8). The largest effect of a 6–7-year cycle is apparent in the ice extent variability of the New Siberian region (from the eastern Laptev Sea to the western East Siberian Sea) and of the Chukchi Sea. The cross-spectral analysis of ice extent for various regions at the frequency corresponding to a 6–7-year cycle showed this variability in the west and east of the Arctic Seas to occur in opposite phase, i.e., their character is close to a standing wave. Unlike this wave, the distribution of the phase difference at the frequency corresponding to a 2–3-year cycle testifies that this variability is manifested in the form of a two-nodal standing wave, with nodal zones passing across the eastern Laptev Sea and along the boundary between the Barents and the Kara Seas.

3

Variability of sea ice thickness and concentration in the twentieth century

3.1 ICE THICKNESS VARIATIONS

Regular measurements of Arctic ice thickness began approximately in the middle of the 1930s in the vicinity of a number of polar stations, some of which were closed in the 1990s. To investigate the changes in landfast ice thickness for this study, we chose 11 stations with approximately equal lengths of observational time series. Five of these stations were located in the Kara Sea (Beliy Island, Dikson Island, Uyedineniya Island, Cape Sterlegov, and Cape Cheluskin), and the others in the Laptev Sea (Tiksi Bay, Kotel'ny Island, Sannikov Island), East-Siberian Sea (Cape Shalaurov, Chetyrekhstolbovoy Island) and in the Chukchi Sea (Wrangel Island).

Figure 3.1 (see color section) plots temporal variability of maximum ice thickness for the period of ice growth and its spectral structure for 1936–2000 based on wavelet-analysis data. Figure 3.1a shows that the maximum thickness of landfast ice in the Kara Sea increased from 1936 to the late 1960s, and then decreased through the end of the twentieth century, but not to its 1936 value. These variations are approximately the same as ice extent variations in the western sector seas (Greenland, Barents, Kara) associated with the "60-year" cycle (Figure 3.1b) discussed in the previous section and are statistically significant in the total wavelet spectrum (Figure 3.1c). The linear trend coefficient for this time series is +0.140 cm/year with a 95% confidence interval of −0.064 . . . +0.344 cm/year. This trend contribution to the interannual variations is only about 3%. The positive trend is noted at four of the five stations in this sea.

The linear trend coefficient for the landfast ice thickness series in the eastern region is very close to zero (Figure 3.1a) and is −0.003 cm/year with a 95% confidence interval of −0.124 . . . +0.119 cm/year; its contribution to interannual variability is negligibly small (0.004%). The linear trend sign is positive at three of the seven eastern region stations and negative at the other four stations. The temporal changes in landfast ice thickness in this region are characterized by the influence of shorter

Table 3.1. Average thickness (cm) of landfast ice during different climate periods in the Arctic Seas, with the anomaly relative to the average value of the observation series in parentheses

Period in years	Kara Sea	Eastern Seas
1936–1957	165 (−8)	199 (0)
1958–1983	181 (+8)	200 (+1)
1984–2000	174 (+1)	197 (−2)

cycles (Figure 3.1b), which are statistically insignificant except for the ~7-year variability (Figure 3.1c).

Table 3.1 presents average values of landfast ice thickness of the "warm" and "cold" epochs for two regions under consideration. Whereas in the Kara Sea the "cold" epoch is distinguished by an insignificantly increased average ice thickness, there are practically no differences between the epochs in the eastern region.

The data presented in the table show that the variations in landfast ice thickness in the Arctic Seas were insignificant during much of the twentieth century. This is also indicated by the analysis of multiyear changes in landfast ice thickness in the Arctic Seas by Buzuyev and Dubovtsev (2002), which does not "confirm climate warming in the area of the Siberian shelf in the 1960s–1990s."

As noted above, the thickness of first-year landfast ice over much of the area of the Arctic Seas comprises 180–200 cm by the end of winter. Karelin (1951) and Nikolaeva and Shesterikov (1970) show that due to the influence of the heat flux from deep Atlantic water, the thickness of drifting ice of the same age in the deep-water part of the Arctic Basin is 15–20 cm less (given the same snow thickness and sums of negative degree-days). The ice growth rate decreases in response to variations in Atlantic water temperature and depth during periods of Arctic warming. Observations made during the drifting expedition of the icebreaker *G. Sedov* indicate that the first-year ice thickness at the end of winter in 1939 was 20% less than that observed during the *Fram* expedition of 1895 in approximately the same region (Buinitsky, 1951).

An analysis of sonar data collected by submarines in the Arctic Basin (Rothrock *et al.*, 1999) showed that the ice thickness there decreased by 1.0–1.5 m, i.e., approximately 40%, from the middle of the 1970s to the beginning of the 1990s. This thinning of the ice cover was attributed to the influence of anthropogenic greenhouse warming. However, analyses of the same data performed by a number of scientists (Shy and Walsh, 1996; McLaren *et al.*, 1994), did not confirm such changes, while others (Wadhams, 1990, 1994) explain this phenomenon by ice ridging at the approaches to Greenland.

The partial concentration of old or multi-year ice (MY) from the ice charts may be used as another proxy for ice thickness data in the Arctic Basin. Though much less accurate than sonar or drilling information, it covers more area and more time intervals. Smolyanitsky (2003) analyzed gridded fields of multi-year partial concentration extracted from the AARI 10-day ice charts in the SIGRID format from the "Global Digital Sea Ice Data Bank" (GDSIDB). The World Meteorological Organization (WMO) Commission on Marine Meteorology (now the Joint WMO/ Intergovernmental Oceanographic Commission for Oceanography and Marine Meteorology, or JCOMM) established the GDSIDB of digital sea ice chart information from the operational ice forecasting centers of participating nations in November 1986. The nominal resolution of a SIGRID grid is 15 minutes latitude. The left column in Figure 3.2 (see color section) shows robust mean MY concentration values for August averaged for 1933–1992 and three sub-periods close to two decades in length: 1940–1959, 1960–1979, 1980–1992. (Air reconnaissance, which ended in 1992, was the prime source of information for AARI ice charts. No calculations were carried out prior to 1940 because of gaps in data taken before that year.) To facilitate interpretation of MY decadal variability, the right column in Figure 3.2 presents differences between three sub-periods and the whole period, or "climatology" from 1933 to 1992. During the first sub-period of the 1940s and 1950s, which at first corresponds to a warmer and then to a colder period in the Arctic, a mixed pattern of MY decrease and increase compared to the climatology is observed in the Eurasian Arctic: decrease in the Kara Sea and the western part of the East Siberian Sea, increase in the northern part of the Barents and Laptev Seas and the eastern parts of the East Siberian and Chukchi Seas. During the second sub-period of the 1960s and 1970s, i.e. during a pronounced colder period in the Arctic, a decrease in MY relative to climatology is observed in the Laptev and eastern part of the East Siberian and Chukchi Seas with an increase in MY relative to climatology in the Kara Sea and the western part of the East Siberian Sea, etc. During the third sub-period, i.e. during a warmer period of the Arctic temperature regime, an increase in MY is observed in the whole East Siberian Sea and the western part of the Chukchi Sea with a decrease in MY in the Barents, Kara and Laptev Seas. It is reasonable to conclude that since the same sign of MY variability is observed in most parts of specific seas, the MY edge also varies in the whole area of seas on decadal time scales and the thermal factor can not be the only cause of its variability.

Recent AARI studies (Gudkovich and Kovalev, 2002a,b) indicate that the sea ice thickness anomalies reported by Rothrock and Maykut (1999) and by Wadhams (1990) are caused by dynamic processes, rather than by thermodynamic processes inherent in global anthropogenic warming. The latter are connected with comparatively short-term changes in atmospheric circulation that control the processes of sea ice advection, ridging, and divergence.

Gudkovich and Klyachkin used a 2-dimensional polynomial to the power of 3 approximation of fields of long-term ice drift vectors, observed by manned stations and automated DARMS buoys for 1937–1975, to study changes in the Arctic ice thickness on annual scales. Their calculations show that the area of decreased end-of-summer ice extent formed in the seas east of Severnaya Zemlya with first- and second-year ice with thicknesses of 1.5–2.5 m, can be transferred as a result of 1–2-year drift to the near-pole area where multiyear ice with a thickness of 3–4 m is usually located.

As a result, a significant decrease in ice thickness is observed, which is then replaced again in 2–3 years by a restored multiyear ice cover typical of this region.

Gudkovich and Guzenko (2007) studied annual changes of the ice thickness in the Eurasian Arctic by tracking displacement of the zone of the former ice cover boundary in the Arctic Basin using ice drift vectors observed by IABP buoys. To that effect resulting vectors of buoy drift recorded for annual intervals were interpolated to positions of the ice cover boundary in order to track its further displacement on annual scale. Figure 3.3 (see color section) presents results of such calculations for a two-year interval from October 1995 to September 1997, which are based on the information from 7 buoys active during October 1995-September 1996 and 9 buoys active during October 1996–September 1997. In spite of significant ice drift anomalies for the period 1995-1997, the result of the calculations confirmed the conclusion above: the boundary of multiyear ice in 1997 was located much farther to the north and east of its mean multiyear location, which explains a decrease in Arctic drifting ice thickness detected in some years by sonar.

Such ice thickness variations depend not only on drift speed anomalies but also to a significant degree on initial ice cover distribution (location of the residual ice edge at the end of summer, boundaries of the ice massifs, etc.). In addition, the alternation of anticyclonic and cyclonic regimes discussed below (Section 4.2) is accompanied by recurring changes in the processes of sea ice convergence and divergence changes. The latter influences sea ice concentration and ridging and hence is responsible for sea ice thickness temporal and spatial variability (Losev *et al.*, 2005; Porubayev, 2000; Makshtas, 2001). Climatic changes can only determine the probability of the formation of the corresponding conditions (initial ice distribution, anomalous character of the baric fields, etc.). It also appears (Makshtas *et al.*, 2002) that the air temperature influences ice thickness but changes in heat fluxes from ocean affected by low-frequency variations in temperature and location of deep Atlantic water in the Arctic Basin play a specific role in ice thickness variations (see Section 4.6).

In order to clarify the problem of the real change in thickness of drifting ice in the Arctic Basin during arctic warming in 2000s, we analyzed the process of ice growth using observations from three Russian drifting stations: NP-32 (2003–2004), NP-33 (2004–2005), and NP-34 (2005–2006). The drift of all three stations was mainly within the near-pole area bounded by parallel 85°N. The data from these stations were made available courtesy of the heads of the stations, V. S. Koshelev, A. A. Visnevsky, and T. V. Petrovsky. It is interesting to compare these data with observations from the drift of the icebreaker *G. Sedov* (1937–1940) during the first twentieth-century Arctic warming. It is important to note that the *G. Sedov* drift in the winter of 1938–1939 was predominantly to the north of 85°N, and hence quite reasonable to compare with the three recent expeditions.

The rate of sea ice growth is known to depend on a number of factors (air temperature, ice thickness and snow cover, their thermal-physical characteristics, etc.). According to Buinitsky (1951), daily ice thickness growth per 1° of average air temperature is determined, at least for first-year ice, predominantly by average ice thickness. Figure 3.4 presents Buinitsky's plot of an empirical dependence using average daily ice growth values per 1° of mean air temperature. These values were

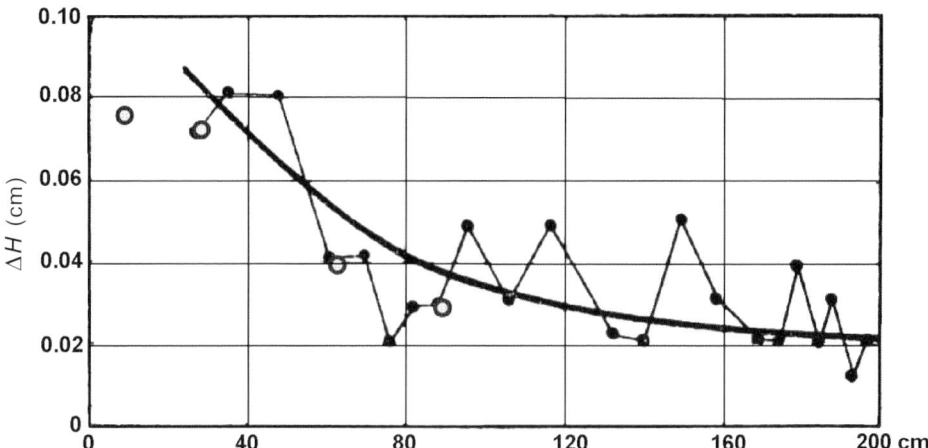

Figure 3.4. The thick black line shows daily ice thickness growth per $1°$ of average air temperature in *G. Sedov* observation data for 1937–1938, plotted by Buinitsky (1951). Gray filled circles denote data from drifting station NP-32 for 2003–2004.

also obtained from intervals of observations of young ice growth at the NP-32 station. These observations were unfortunately interrupted due to breakup of the station's ice floe at the beginning of January 2004.

The values shown confirm the general character of the dependence. However, they were 0.01 cm/(°K · day) lower, on average, compared to the empirical curve, which corresponds to an ice growth deficit of 7 cm/month. This is explained to a great extent by increased average snow cover thickness (23 cm at the NP-32 station compared to 6 cm in the data of the expedition onboard the *G. Sedov*). Calculations of residual (first- and second-year) ice growth require accounting for snow cover depth as well as ice thickness. The heat conductivity coefficient of snow (0.3 W/(m · °K) is known to be 7 times less than that of ice (2.2 W/(m · °K) (*Sea ice*, 1997). That is why snow cover with a thickness of 0.2 m (for example) has the same influence on the growth of 1 m thick ice as 2.4 m thick ice. This is quite accurately accounted for in the following equation (Nikolayeva and Shesterikov, 1970; Frolov *et al.*, 2005):

$$H = -7.0h + \sqrt{(7.0h + H_0)^2 + 0.00122(\Theta - T_s)\tau - Fw \cdot \tau/(L \cdot \rho)}, \qquad (3.1)$$

where H_0 and H are the initial and final ice thickness, respectively; h is snow thickness; T_s is average snow surface temperature; Θ is the freezing temperature of water near the lower ice surface; and τ is the time interval. Here, ice and snow thickness are expressed in meters and τ in days. The third term is responsible for the influence of heat flux from water, where Fw is heat flux from water, and L and ρ are the heat of melting and the density of ice, respectively.

Table 3.2 presents the initial data and the results of ice thickness calculations for monthly time intervals using Equation 3.1. The calculations were based on data and

Table 3.2. Observation data on the ice thickness growth (m) during the icebreaker G. *Sedov* expedition and at the NP-32, NP-33, and NP-34 drifting stations, and the results of calculations using these data

Expedition	Year	Months	T^0_{avg}	H_0	H	h_{avg}	δH^*	Notes
G. Sedov	1937/38	XI–V	−20.9	0.33	1.95	0.06	+0.04	First-year
G. Sedov	1938/39	XI–V	−26.5	0.64	2.05	0.19	+0.06	First-year
G. Sedov	1938/39	XII–V	−27.3	0.99	2.02	0.23	−0.04	First-year
G. Sedov	1938/39	XII–V	−27.3	1.46	2.16	0.31	−0.01	Second-year
NP-32	2003/04	IX–XII	−19.7	0.00	1.05	0.23	+0.07	First-year
NP-32	2003/04	X–II	−29.8	1.55	1.92	0.33	−0.04	Second-year
NP-33	2004/05	XI–V	−25.0	1.97	2.42	0.52	−0.01	Second-year
NP-34	2005/06	X–IV	−22.3	1.00	1.92	0.31	+ 0.02	First-year

* δH is the average difference between the observed and the calculated ice thickness for the end of the calculation month.

information from the G. *Sedov* expedition and the NP-32, NP-33, and NP-34 drifting stations. The value of Θ is assumed to be equal to $-1.7°C$ (Anon. (F)). Instead of T_s values, air temperature T_a was used. The differences between them are noticeable at ice thicknesses up to 0.5 m with an insignificant snow cover.

As Table 3.2 shows, the relative error in ice thickness calculations at the end of winter is only 0.5–3%. If we assume that the average excess of the calculated ice thickness values over the measured values in winter of 2003–2004, when the NP-32 station was closer to Fram Strait, is determined by the influence of the heat flux from deep Atlantic water, then it is simple to estimate the value of this flux using the method described above. The calculations show that total heat flux from the ocean for the winter could not be more than $100{,}000 \, kJ/m^2$ (about $2.5 \, kcal/cm^2$). According to Panov and Shpaikher (1963), the maximum value of this flux near the continental slope of the Siberian Arctic Seas is $5–6 \, kcal/cm^2$, and from better documented estimates by Nikolayeva and Shesterikov (1970), $4.0 \, kcal/cm^2$ (up to $1.0–1.2 \, kcal/cm^2$ in the deep-water part of the Arctic Basin).

In addition to heat from the ocean, an important factor slowing ice growth is the presence of melt-ponds. Their full-depth freezing delays the beginning of growth on the bottom surface of the ice cover, sometimes until the beginning to the middle of December. According to observations at the drifting stations, the area of the melt-ponds by the beginning of freeze-up comprised up to 40% of the surface of the ice which survived the summer melt. Therefore, and given the absence of significant error in calculated ice thickness, it can be considered that the influence on the ice growth of heat flux from the water during the epochs under consideration (1937–1939 and 2003–2006) was not significant. It is likely that in the continental slope zone to the north of the Eurasian Arctic Seas shelf, where the main flow of deep Atlantic water is located, the influence of this heat on ice growth is more obvious than in the near-pole area.

As the data in Table 3.2 show, average wintertime air temperatures during the first and second warming in the Arctic differ insignificantly, and there is no significant difference in ice thickness at the end of winter. Note again that according to Vize (1951) and Buinitsky (1951), winter air temperature during the drift of the icebreaking vessel *G. Sedov* was 6–8 degrees higher than at the time of the *Fram* drift. The sum of negative degree-days at the end of the 1930s, similar to the beginning of the twenty-first century, was 21% less than at the end of the nineteenth century. Ice growth in the near-pole area has decreased by exactly the same value (Karelin, 1951). So, the results presented above contradict estimates of catastrophic (almost twofold) ice thickness decrease in the Arctic Basin during the last decades of the twentieth century (Shimada *et al.* (2006), Anon. (A), Anon. (B) and Anon. (C)). It should be noted that in the Vize (1951) and Buinitsky (1951) data, the mean monthly air temperatures in spring-summer of *G. Sedov* drift were 0.2–0.5°C lower than at the time of *Fram* drift, which indicates the secondary importance of ice melting compared to ice growth in ice thickness variations in the Arctic Basin.

3.2   CHANGES IN ICE CONCENTRATION

Along with the ice extent and thickness, an important characteristic of the ice cover is its concentration. Ice cover concentration is an important factor for ice navigation and in the exchange of energy between the ocean and the atmosphere.

Zakharov (1996) shows that ice concentration in the Arctic Seas changes significantly during the summer season. In the basin itself, changes in ice concentration are not greater than fractions of a percent, on average (Vowinchel and Orvig, 1973), although in some limited areas they can increase significantly in some years. Specific characteristics of different concentrations of ice area in the Arctic Seas and their causes are considered in Gudkovich and Zakharov (1998).

Ice concentration in these seas varies significantly with climatic changes. Zakharov (1996) presents charts showing average changes in ice cover concentration at the beginning of September from the last decade of Arctic warming (1930-1940s) through the following cooling period. These charts show a significant increase in ice cover concentration in the northeastern Kara Sea during the cooling epoch.

Smolyanitsky (2003) used the same approach for gridded fields of sea ice total concentration as that described in Section 3.1 for MY partial concentration. Figure 3.5 (see color section) charts the changes in average ice cover concentration in the Arctic Seas from the 1940–1962 warm epoch to 1963–1983 cold epoch.

The upper part of the figure characterizes the first half of June, when the changes in the Barents Sea are better expressed, and the lower part characterizes the middle of August, when the changes are more pronounced in the Siberian shelf seas. As the figure shows, there was a significant increase (more than 4–5 tenths) in ice cover concentration from the warm to the cold epoch in the northeastern areas of the

Barents and Kara Seas, in the Pechora Sea, and also in the central East Siberian Sea and to the north of the Chukchi Sea (more than 3–4 tenths). In the Laptev Sea, a small decrease in average concentration is noted, which may be connected with elevated ice export from this sea during cold epochs as compared with the warm epochs (see section 4.4).

4

Consistency among sea ice extent and atmospheric and hydrospheric processes

The variability and state of the Arctic sea ice cover strongly depend on atmospheric conditions as well as ocean dynamic and thermodynamic processes (Alekseev, 1976; Appel and Gudkovich, 1992; Gudkovich *et al.*, 1972; Doronin, 1969; Doronin and Kheisin, 1975; Zakharov, 1981; Zubov, 1938, 1944; Shuleikin, 1953; Wadhams, 1994). A number of parameters influence the direction and intensity of these processes. The most significant are: the surface air temperature, wind, oceanic boundary layers and their stratification, and ocean circulation.

In order to understand the causes of long-term changes in the ice cover, it is necessary to define the temporal and spatial relationships of the sea ice cover with all the factors mentioned above. This section analyzes the connections of various large-scale processes to climatic changes, as considered in Gudkovich and Kovalev (1997).

4.1 LONG-TERM CHANGES IN ARCTIC AIR TEMPERATURE

Anomalies of mean annual surface air temperature (SAT) in the zone from 70–85°N for the period from 1900 to 2003 were used to analyze climatic changes observed in the Arctic Seas throughout the last century. The anomalies were calculated on the basis of the archive of mean monthly air temperature in a grid consisting of cells (5° of latitude × 10° of longitude) drawn in the area from 20°N to 85°N for 1891–2000. These data are based on the SAT charts of the Northern Hemisphere constructed at the Main Geophysical Observatory in St. Petersburg, Russia, using all known data published in various climatologic summaries. In the late 1970s, the air temperature data were digitized by the USSR Hydrometeorological Center and updated using Hydrometeorological Center and AARI Department of Long-Range Weather Forecasting data.

Periodic cooling and warming events are evident in air temperature fluctuations in the Arctic during the twentieth century, similar to the changes in the ice cover

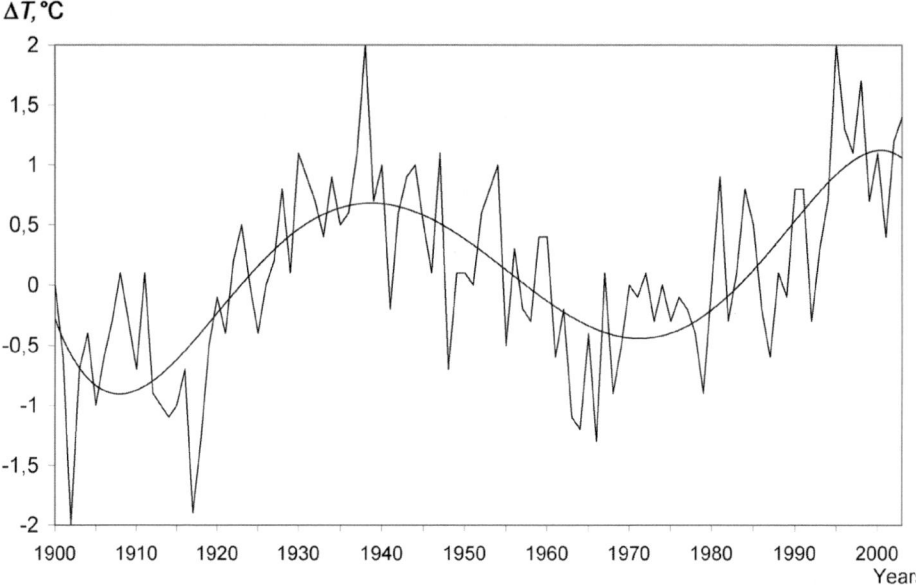

Figure 4.1. Changes in mean annual air temperature anomalies (ΔT) in °C in the 70–85°N zone in the twentieth and early twenty-first centuries and their polynomial trend.

discussed above (Figure 4.1). A cool period at the beginning of the century was replaced by warming in the 1920s–1940s that is referred to in climatic literature as the "Arctic warming period." Then, a relatively cooling trend was observed from the late 1950s to the late 1970s, which was in turn replaced by a new warming trend at the end of the century when the temperature reached its maximum in the late 1990s to the early 2000s. The trend seen in Figure 4.1 approximated by a polynomial to the sixth power, suggests that the duration of this cycle is about 50–60 years.

The interpretation of a 50–60 year cycle as the main climatic fluctuation in the Arctic in the twentieth century is also supported by the wavelet-spectrum of mean annual air temperature anomalies, from which a linear trend was deduced (Figure 4.2, see color section). Figure 4.2 quite clearly shows the main features of surface air temperature variability in the high-latitude zone with alternation of cold and warm phases in a 60-year cycle.

Fluctuations of mean air temperature in the Northern Hemisphere (Minobe, 1997, Klyashtorin and Lyubushin, 2004), the Earth's rotation speed (Rudyaev *et al.*, 1985), ice export from the Arctic Basin (Gudkovich *et al.*, 2007), and other indicators also reflect this cycle. Their global nature is apparent in paleoclimate data from ice cores collected at Vostok station in Antarctica, which were analyzed for the isotopic composition of atmospheric precipitation for the last 200 years. (Lipenkov *et al.*, 2002, 2003; Figure 4.3).

Figure 4.1 shows the changes in mean annual twentieth century air temperature anomalies in the high-latitude zone of the Northern Hemisphere as well as "60-year"

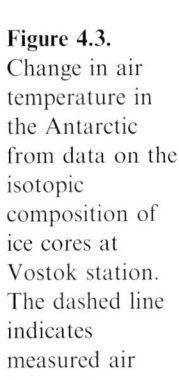

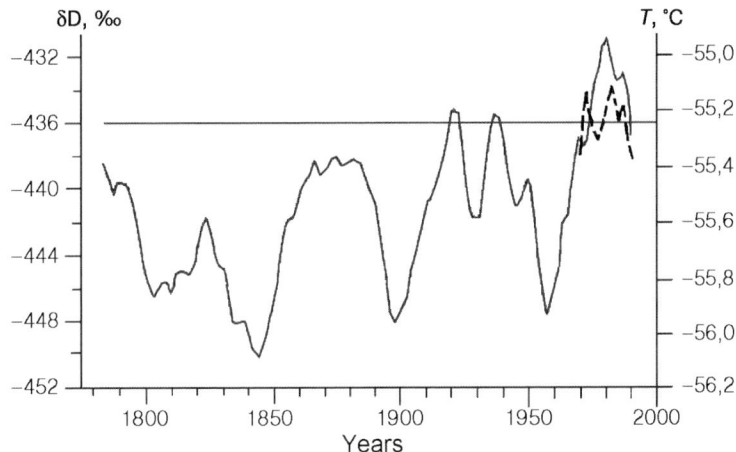

Figure 4.3.
Change in air
temperature in
the Antarctic
from data on the
isotopic
composition of
ice cores at
Vostok station.
The dashed line
indicates
measured air
temperature.

cycles of interannual fluctuations within ±2°C. These fluctuations occurred against
the background of an even longer change, a positive linear trend showing a gradual
increase of 0.8–0.9° in Arctic air temperature during the century. This trend may fit
into one of the multi-century climate fluctuations that have been observed in Earth's
history (Monin and Sonechkin, 2005). Calculations of the wavelet-spectrum of a 400-
year series of reconstructed anomalies of mean annual air temperature in the region
from 17.5°N to 87.5°N for the period 1579–1983 (data taken from Bashkirtsev and
Mashnich, 2004) show significant peaks at the frequencies corresponding to 200 and
100 years (Figure 4.4, see color section).

A stable cycle with an average duration of about 210 years is also found in data
on beryllium-10 isotope concentration (responding to air temperature changes)
contained in dendrologic evidence from the northern Eurasian forestry boundary
(Raspopov *et al.*, 2004). It is possible that part of this cycle contributes to the linear
trend in twentieth century air temperature.

An analysis of consistency among the main components of SAT changes in
the Arctic and in the hemisphere in general is of great interest. The correlation
coefficients characterizing this consistency are quite large: 0.59 (1900–2003) and
0.70 (1971–2003). Table 4.1 compares the characteristics of the linear trends and
the "50–60-year" fluctuations in three geographical areas of the Northern
Hemisphere in the twentieth century. The assessment methodology is similar to
that used in Sections 2.2 and 2.3. The table shows much greater variability in air
temperature and its two main climatic components in the Arctic than in the
temperate latitudes and over much of the Northern Hemisphere. The contribution
of the linear trend of mean annual temperature, averaged over the corresponding
area, increases with decreasing latitude, while the "50–60-year" cycle mean annual
temperature decreases. The possible causes of these changes will be considered in
Section 5.4.

The fact that weather and climate variability increases with latitude is known as
"polar amplification." A model proposed by Alekseev and Svyashchennikov (1991)

Table 4.1. Characteristics of mean annual twentieth century air temperature variation in three zones.

Region	rms	Trend coefficient (deg/year)	Average cycle amplitude	Trend contribution to dispersion	"50–60-year" cycle contribution to variance
70–85°N	0.78°	0.0097	0.65°	13%	39%
40–65°N	0.34°	0.0048	0.25°	17%	27%
17.5–87.5°N	0.26°	0.0069	0.17°	59%	23%

explains this phenomenon by taking into account heat advection in the atmosphere that results in air mixing between adjoining latitudinal zones. Zakharov (1996) examined the relationship between the maximum air temperature variability zone and the location of the frontal area between the Arctic and the marine polar air masses. This allowed him to conclude that polar amplification is a simple result of the mobility of the polar front and its fluctuations in time, and furthermore, he was also able to explain localization of the most significant climatic changes in the sub-Atlantic Arctic. For most of the year, the mobile ice edge is located in this region, where the Arctic front passes between 70° and 80°N, and the horizontal temperature gradients are most pronounced near it. However, recognizing the important role of the North Atlantic in generating low-frequency climate fluctuations, Polyakov et al. (2002) express doubts that the observed air temperature trends in the Arctic confirm the hypotheses of polar forcing of global warming.

Alekseyev et al. (2004) show convincingly that seasonal twentieth century climate changes occurred extremely irregularly over the Earth's surface. The spatial non-uniformity of air temperature changes was connected with both geographical latitude and the longitude of the region. It is important to note that a negative correlation was revealed in this study between the mean zonal air temperatures at high and middle latitudes in some seasons.

Increased air temperature variability in temperate latitudes of the Eurasian and North American continents is typical of the winter months. The signs of temperature anomalies over the oceans and the continents are usually opposing. Such temperature field structures were called COWL (cold ocean warm land) by Wallace et al. (1995), who accurately related these phenomena to increased west-to-east transfers in the atmosphere during warming periods and their attenuation during cooling periods.

Klimenko (2007) presents similar findings. He charts the differences between mean annual and seasonal air temperatures in the Northern Hemisphere during the warmest 20-year period (1986–2005) and the coldest 20-year period (1911–1930) and shows the maximum warming to cover the temperate latitudes of Eurasia and North America. The charted differences in mean annual temperatures at the warming epicenters exceed 1.5°C (5°C in the winter season); that is, a tenfold increase in the average global signal. This is much greater than warming in the Arctic region adjoining the North Atlantic; however, a stricter approach to estimating the values under consideration requires a preliminary exclusion of cyclic fluctuations.

Klimenko (2007) reports that no warming was observed during the same period in the northern parts of the Atlantic and Pacific Oceans. Hassol (2004) shows most of the North Atlantic in the zone of decreased (by 1°C) mean annual and winter air temperature during 1954–2003. There is further evidence in 3,000 years of data on surface temperatures in the Sargasso Sea based on the oxygen ratio in the remains of plankton organisms buried in bottom sediments (Sorokhtin, 2001, Keigwin, 1996).

A direct cause of the patterns noted is, undoubtedly, intensification of zonal (west-to-east) transfers in the atmosphere of temperate latitudes during periods of climate warming, as discussed below in Sections 4.2 and 4.7. A corresponding increase in heat advection from the oceans to the continents plays an important role, as does moisture advection, which is accompanied by increased cloudiness, resulting in the increase of both long-wave counter-radiation in the atmosphere and temperatures in the lower atmosphere. As expected, loss of heat from upper ocean layers, especially in winter, is confirmed by the data presented in the aforementioned studies.

In addition to a positive trend and a "50-year" SAT cycle, the Arctic SAT exhibits significant interannual variations. Figure 4.5 shows that when the linear trend and the "60-year" cycle are excluded from the mean annual temperature anomalies considered above, the largest contribution to these variations comes from relatively high-frequency, variable-amplitude, cyclic fluctuations lasting 2–3 years; these are overlain by longer cyclic fluctuations, which are approximately represented by 5-year running averages. A spectral analysis of this curve reveals the dominance of a "20-year" cycle (Figure 4.6, see color section).

A periodogram analysis of the data smoothed by 5-year periods allowed us to estimate the amplitude of the "20-year" cycle and its contribution of about 5% to the variance in Arctic air temperature. The data summarized in Table 4.1 indicate that

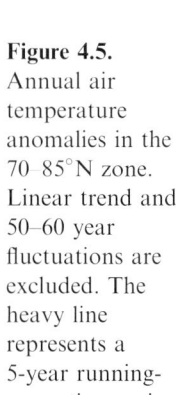
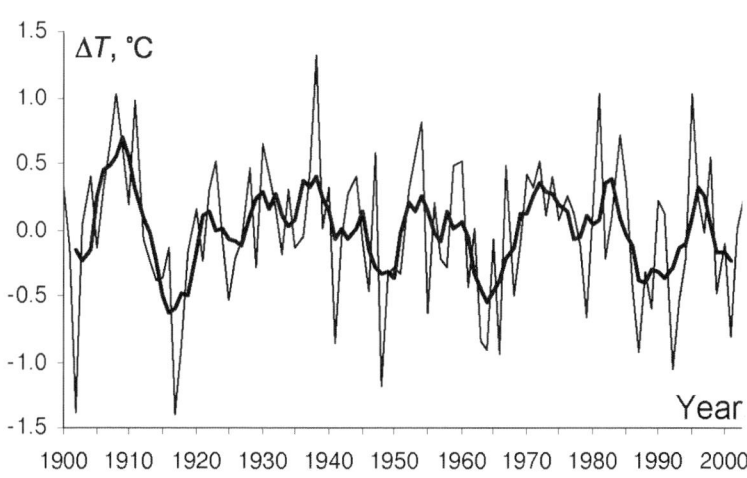

Figure 4.5. Annual air temperature anomalies in the 70–85°N zone. Linear trend and 50–60 year fluctuations are excluded. The heavy line represents a 5-year running-mean time series.

the mean annual Arctic SAT changes appear to be mainly due to fluctuations lasting more than a century (a linear trend) as well as 50–60 year and 20-year cycles. Their total contribution to the variance comprises 57%, and hence the contribution of relatively high-frequency fluctuations is 43%.

4.2 LONG-TERM CHANGES IN ATMOSPHERIC PRESSURE FIELDS AND ATMOSPHERIC CIRCULATION INDICES

As in other hydrometeorological fields, study of changes in sea level pressure (SLP) fields presents major problems related to the complicated spatial distribution of the changes in time. The simplest method of investigating the characteristics of the fields in different climatic epochs is to calculate the average fields with a subsequent calculation of their differences. Korolev and Subbotin (1988) investigated the characteristics of the SLP and SAT fields in the Northern Hemisphere during the Arctic "warming" and "cooling" periods using this method. The differences in the average fields over 24 years for each period allowed them to identify deepening of the Icelandic and Aleutian winter depressions during the "warming" epoch as compared with the "cooling" epoch. They found that zones of maximum differences in SAT do not coincide with regions of the maximum differences in atmospheric pressure.

A similar method was used by Gudkovich et al. (1994, 1997) to investigate the characteristics of SLP, SAT, surface water salinity, and other climate indicators during the epochs of increased and decreased ice cover area for the North European Basin, determined by 20-year fluctuations. The results indicate that the distribution of SLP anomalies during the "cold" epochs contributes to increased ice export from the Arctic Basin to the Greenland and Barents Seas, and decreased ice export to the Arctic Basin from the Kara Sea as compared with the "warm" epochs. This is consistent with conclusions reported by Subbotin (1988). Significant surface water freshening in the Barents Sea and especially in the Kara Sea also occurs during the "cold" epochs, whereas during the "warm" epochs, the surface water salinity noticeably increases. This is connected to some degree with corresponding river runoff fluctuations, but mainly with the changed inflow of relatively saline water of Atlantic origin (Appel and Gudkovich, 1984).

The atmospheric pressure difference between Franz-Josef Land and Cape Zhelaniya of Novaya Zemlya is caused by the intensity of northeasterly (or southwesterly) winds and indicates the anomalous character of Barents Sea water inflow to the Kara Sea in this region. This relationship is confirmed by data from the annual series of ocean currents measured by moorings along a transect described by Loeng et al. (1993). The atmospheric pressure difference between Wrangel Island and Cape Barrow provides an indicator of Bering Sea water inflow to the eastern Arctic Seas. An increase in this difference indicates greater incidence of prevailing northeasterly winds, resulting in intensification of the Long Strait branch of the Bering Sea current that brings relatively saline water to Long Strait; a decrease in the pressure difference (or a change of its sign) is accompanied by the opposite effect (Gudkovich et al., 1972). This is confirmed by an empirical relationship between ice conditions in the southwestern Chukchi Sea and corresponding indicators of air transfers (Santsevich *et al.*, 1979), and by the interrelationship of surface water salinity near the Chukchi coast and the pressure difference along the transect mentioned above.

These pressure differences work together, affecting ice exchange, ice growth, and melting in the sub-Atlantic Arctic Seas, resulting in significant changes in the ice cover area in this region. The response of the atmosphere to ice cover changes, along with the relative stability of the thermohaline water structure, contributes in turn to persistence of corresponding anomalies for several years, causing fluctuations in the state of the atmosphere, the ocean, and ice cover. In the seas that are remote from the Atlantic Ocean, there are no conditions that result in the collective impact of many factors; thus, relatively short-term interannual variations of sea ice extent predominate here, and comparatively long climatic changes are weaker.

A disadvantage of the "differences" method of investigating the variability of the hydrometeorological fields is the necessity to prescribe the duration and phase of cyclic changes. Due to phase non-coincidence (in space and time) of different hydro-meteorological characteristics, it is difficult to achieve unambiguous results because different cycles influence the anomaly of the characteristic for a specific period.

This can be partly avoided by expanding the study fields into empirical orthogonal functions (EOFs) (Bagrov *et al.*, 1959). Applying this method makes it possible to calculate multidimensional field expansion vectors, thus characterizing the spatial structure of their variability, and to calculate time coefficients, thus describing a change in time for each vector. The advantage of the method is noise filtration (high-frequency changes) and information compression for describing a complex of fields.

Applying this method, the first several (3–5) vectors of the F field expansion describe a significant (up to 80%) fraction of variability. A spectral analysis of the temporal coefficients of different components allows us, in some cases, to reveal spatial characteristics of different cycles (e.g., Baranov *et al.*, 1986; Vangengeim, 1986; Baranov and Vangengein, 1988; Korolev and Subbotin, 1988). However, in order to get statistically robust results, application of the F method is restricted by the conditions of data series length, size and location of the area of the expansion components.

For characterizing general atmospheric circulation processes, different quantitative indices are often used for investigating the intensity of atmospheric circulation, especially its zonal and meridional components. In the past, the indices enumerated by Rossby, Blinova, Vittels, Belinsky, and Kats were used (see Girs, 1960). The indices also include recurrence of general types (forms) of atmospheric circulation as described by Vangengeim (1935) for the Atlantic-European sector of the Northern Hemisphere: western (W), eastern (E) and meridional (C) circulation descriptions that were further developed by Girs (1960) for the Pacific Ocean sector of the hemisphere (circulation types Z, M_1, and M_2). The Vangengeim–Girs system of classification has been widely used for developing long-range meteorological forecasting at the AARI and other research institutes.

Karklin *et al.*, (2001) reveal the interrelationship of long-term changes in sea ice extent and atmospheric processes by employing the Vangengeim–Girs generalized indices of atmospheric circulation. Examining the cyclic changes in the number of days with atmospheric circulation forms C and $W + E$ occurring in opposite phase during the twentieth century, the authors identify cycles lasting about 60 years. The phases of the $W + E$ fluctuations and the total sea ice extent of the Siberian shelf seas practically coincide; however, unlike sea ice extent whose changes exhibit a negative trend, no trend was apparent in fluctuations of atmospheric circulation forms.

The presence of a 50–60 year cycle in changes in the atmospheric circulation indices is confirmed by studies of Klyashtorin and Lyubushin (2004). Their work reveals the close relationship of the changes in these indices and the cyclic changes in global air temperature anomalies, the level of Lake Balkhash (120-year observation series), maximum levels at the Neva River mouth (more than 120 years), precipitation on the West Coast of North America (more than 100 years), and Barents Sea ice extent (about 100 years). All of these parameters exhibit prominent cyclic changes with periods of 55–60 years. Except for changes in the anomalies of global air temperature and sea ice extent in the Barents Sea, which include a noticeable linear trend, the other indicators under consideration lack the linear trend similar to mean annual air temperature anomalies in the 60–85°N zone. This work also indicates that recurrence of atmospheric circulation forms $W + E$ correlates with climatic indices of the Pacific Ocean Decadal Oscillation (PDO) and the Aleutian low of atmospheric pressure (ALPI), which also include 60-year cycles.

Klyashtorin and Lyubushin (2004) demonstrate a close connection between atmospheric circulation indices and other climatic indicators (i.e., anomalies of global surface air temperature; water temperature in the 200-m layer at the Kola meridian, $33°30'E$; air temperature on Jan-Mayen Island) and biomass concentration (herring and cod) in North European Basin waters. Similar 55–60-year cycles were also detected in the long-term changes in this biomass.

In addition to a 50-year cycle, cycles of 20–25 years and about 10 years were revealed in the changes in sea ice extent of the Arctic Seas, as shown above. The 20–25-year cycles occur in various ocean processes, in the ice cover, and in the atmosphere connected with the phenomenon of self-oscillations in the Norwegian energy-active zone of the ocean (NEAZO) (Gudkovich and Kovalev, 2002a) (see Section 5.3). The magnetic cycle of solar activity and the declination cycle of tide inequality are also close to 20 years.

The AARI uses an index of high-latitude zonality (I_z) proposed by Dmitriyev (1994, 2000) for scientific and operational work. This index characterizes an average geopotential difference at the 500 hPa surface between the parallels 60°N and 80°N and hence reflects the intensity of zonal transfers in the atmosphere of high northern latitudes. Figure 4.7 shows changes in I_z-index anomalies smoothed by 11-year periods and averaged for April–October during the second half of the twentieth century. Twenty-year zonal-flow fluctuations stand out clearly against the background of a noticeable linear trend, indicating the gradual intensification of a cyclonic vortex over the Arctic. The linear trend's contribution to the total variance of I_z, estimated by Equation 2.1, comprises 12.6%, and the contribution of a 20-year

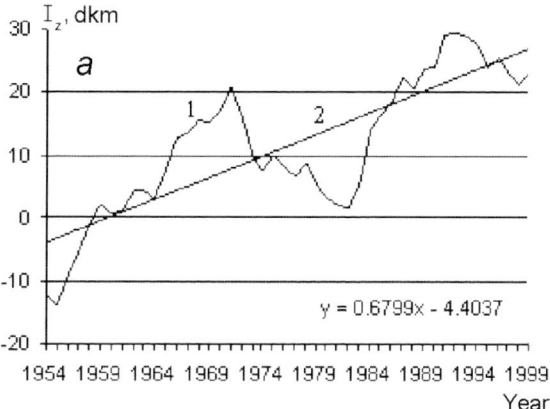

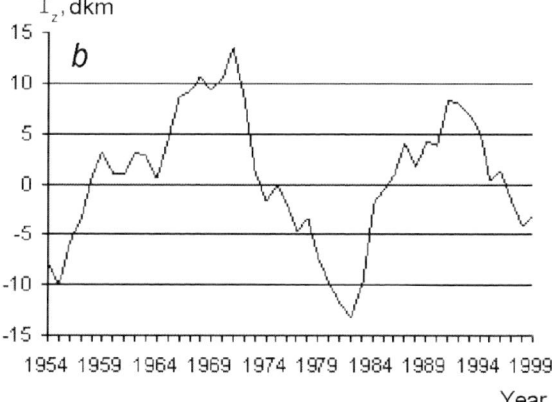

Figure 4.7. (a) Indices of high-latitude zonality smoothed by 11-year periods (1) and averaged for the warmer part of the year (April–October). (b) The same with the linear trend excluded (2).

cycle, determined by Equation 2.2, comprises 28.3%. The effect of smoothing was taken into account in determining the average amplitude.

There is also a 10-year cycle in changes in the high-latitude zonality index. It was first observed in the early 1960s in changes in the Arctic Basin water circulation system that concern both the size and location of the Beaufort anticyclonic circulation and the location of the Transarctic Current core. The average period of these fluctuations was 8–10 years. Regime types A (anticyclonic) and B (cyclonic) revealed in data from oceanographic surveys of the Arctic Basin are distinguished by the characteristics of hydrometeorological and ice conditions of the Arctic Seas, including the location of the trans-polar drift core and dimensions of the Beaufort gyre (Gudkovich, 1961).

Recent analyses of wind-driven circulation in the Arctic Ocean by Proshutinsky and Johnson (1997), Polyakov, Proshutinsky and Johnson (1999) and Proshutinsky *et al.* (1999) show that wind-driven ice motion and upper ocean circulation alternate between anticyclonic and cyclonic regimes. Shifts between regimes occur at 5-year to

7-year intervals, resulting in 10-year to 15-year periods. The anticyclonic circulation regime is evident in the model results for 1946–1952, 1958–1962, 1972–1979, 1984–1988, and 1997–present. The cyclonic circulation regime prevails in the results for 1953–1957, 1963–1971, 1980–1983, and 1989–1996. Based on these analyses, these authors proposed an Arctic Ocean Oscillation (AOO) index showing alternation of the cyclonic and anticyclonic regimes.

Confirming the Proshutinsky and Johnson (1997) theory, Thompson and Wallace (1998) analyzed SLP fields for latitudes higher than 15°N using EOF analysis, introduced an index of the first EOF mode of SLP, and named it Arctic Oscillation (AO). The AO can be characterized as an exchange of atmospheric mass between the Arctic Ocean and the surrounding zonal ring centered at ~45°N. The observed trend in the AO (Thompson and Wallace, 1998; Rigor et al., 2002) toward its high index polarity (i.e., toward stronger westerlies at subpolar latitudes and lower SLP over the Arctic) is a way of interpreting the observed decrease in SLP over the North Pole and the associated cyclonic tendency in surface winds over the Arctic (Rigor et al., 2002); this is similar to Proshutinsky and Johnson's (1997) description of the cyclonic circulation regime. The most valuable part of the Rigor et al. work is that, based on observational sea ice drift data, they show direct responses of arctic surface circulation to wind forcing, and they show that changing sea ice conditions depend on atmospheric conditions related to the AO index. It was shown by Proshutinsky et al. (1999) that the AO phenomenon expressed by this index includes both 10-year and 20-year components.

In order to assess the influence of the Arctic Oscillation on sea ice extent in the main regions of the Arctic Ocean—the North European Basin and the Siberian Arctic Seas—we calculated the average ice cover areas in August in anticyclonic and cyclonic regime years. It turned out that the differences in sea ice extent in both regions for the indicated groups of years are practically absent. However, it is of interest that changes in the ice area from the beginning to the end of each cycle for the same groups of years differ quite significantly. To exclude the influence of short-period fluctuations presented in Table 4.2, the sea ice extent changes were subjected to 5-year smoothing.

Table 4.2 shows that in 88% of cases during anticyclonic regimes, sea ice extent increases in the North European Basin and decreases in the Siberian Arctic Seas, while cyclonic circulation has the opposite effect. The absolute value of changes in the Siberian Arctic Seas is more than 5 times higher than in the North European Basin. Such character of sea-ice extent dependence on the anomalies of the circulation regime indicates an accumulation (integration) of impacts of the latter on sea-ice extent during each cycle. As a result, the sea-ice extent fluctuations are displaced in phase from the corresponding fluctuations in atmospheric circulation by 1/4 of the period. The lag in sea ice-extent changes compared to variations in the AOO indices may also be caused by a lag in the formation of real baroclinic currents from the calculated barotropic components changing in response to changes in the wind fields.

The high-latitude zonality index (I_z) and AOO phases are interrelated: the average index value is negative in years with an anticyclonic AOO regime and positive in the years with a cyclonic regime (Table 4.3).

Table 4.2. Changes in the ice cover area in August from the beginning to the end of the circulation cycles in Arctic Ocean regions (in 10^3 km^2)

Circulation regime	Years	North European	Siberian Arctic
Anticyclonic	1946–1952	+5	−259
	1958–1962	+44	−24
	1972–1979	+60	−87
	1984–1988	+23	−308
	Average	*+34.5*	*−170*
Cyclonic	1953–1957	−37	+209
	1963–1971	−52	+144
	1980–1983	+36	+39
	1989–1997	−4	−10
	Average	*−14.2*	*+95*

Table 4.3. Average high-latitude zonality index I_z values (in decameters) for anticyclonic and cyclonic regimes (1949–1997)

Period	Circulation regime		Multiyear average
	Anticyclonic	Cyclonic	
May–October	−3.3	+17.9	+8.4
January–December	−4.2	+15.8	+5.0

On average, the high-latitude zonality index reflects the differences between atmospheric circulation in the Arctic typical of years with cyclonic and anticyclonic regimes (Figure 4.8).

To characterize the intensity of the west-to-east transfer in the atmosphere of temperate latitudes, Blinova (1943) proposed an index of average geopotential difference at the 500 hPa surface between 40°N and 50°N and 55°N and 65°N. Using the Blinova (1943) method, we calculated a zonality index on the basis of mean monthly SLP for all months of each year from 1900 to 2000. Figure 4.9 presents the anomalies of mean annual values of this index.

There are significant interannual fluctuations in the zonality index. However, in the course of multiyear index variability, some tendencies stand out in periods of intensified and weakened west-to-east transfer in the atmosphere of temperate latitudes. These trends are clearly identified as a result of approximating the index values using a polynomial to the power of 6 as shown in Figure 4.9, which reveals a prominent "60-year" cycle in changes in different hydrometeorological indicators of Earth's climatic system.

Although the index reflects the variability of zonal circulation at temperate latitudes, there is a noticeable synchronicity in the trends of its variability with climatic changes occurring in the Arctic latitudes during the twentieth century. The average index values increased during the periods of Arctic warming in 1920–1940 (the average index anomaly is +0.4 hPa) and in the 1980s–2000s (the average

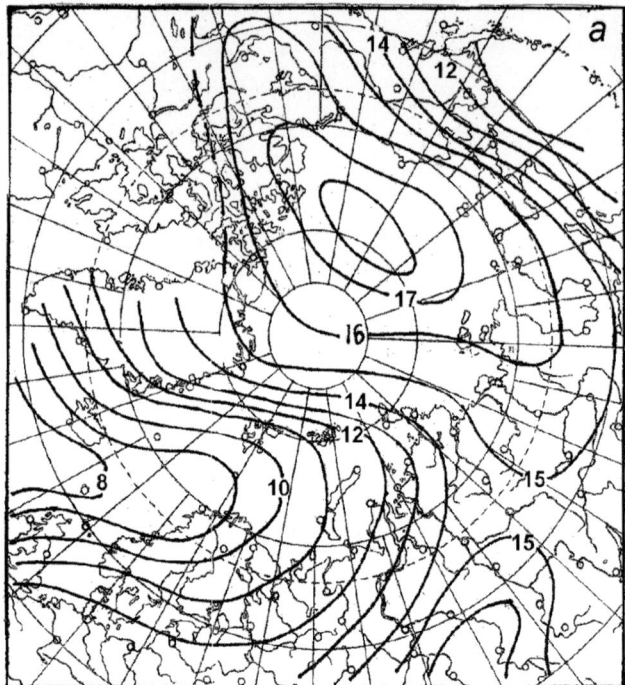

Figure 4.8. Atmospheric pressure distribution at sea level averaged for periods of anticyclonic (a) and cyclonic (b, facing page) circulation regimes, and the differences between them (c, facing page).

anomaly is $+1.0$ hPa), and they decreased during the cold period 1955–1975 (the average anomaly is -1.0 hPa). This indicates that climatic changes occurring at high and temperate latitudes of the Northern Hemisphere are related. Judging from the data available, zonal processes in the hemisphere were more weakly developed during the first Arctic warming (1920–1940) compared to the second warming.

Changes in the zonality index result from atmospheric circulation modification in different climatic periods, which is manifested in the SLP variability of high and temperate latitudes.

In Figure 4.10, the fields of mean annual SLP anomalies averaged for 10-year periods of warming and cooling in the Arctic are compared (1990–2000 and 1965–1975, respectively). These 10-year periods were chosen near the extremes of the 60-year cycle of air temperature fluctuations in the Arctic. In the "cold" years, the atmospheric pressure at polar latitudes and within a significant part of temperate latitudes increases (Figure 4.10a), and in the warm years, it decreases (Figure 4.10b). It should be noted that subtropical latitudes are characterized by the opposite sign in atmospheric pressure changes.

As noted earlier (Korolev and Subbotin, 1988), the Icelandic and Aleutian depressions grow deeper from the cooling period to the warming period, while the Siberian and Arctic atmospheric pressure highs become weaker (Figure 4.10c). The cause of this atmospheric modification is intensified cyclonic activity in the Arctic, which affects the expansion and deepening of the circumpolar vortex in the

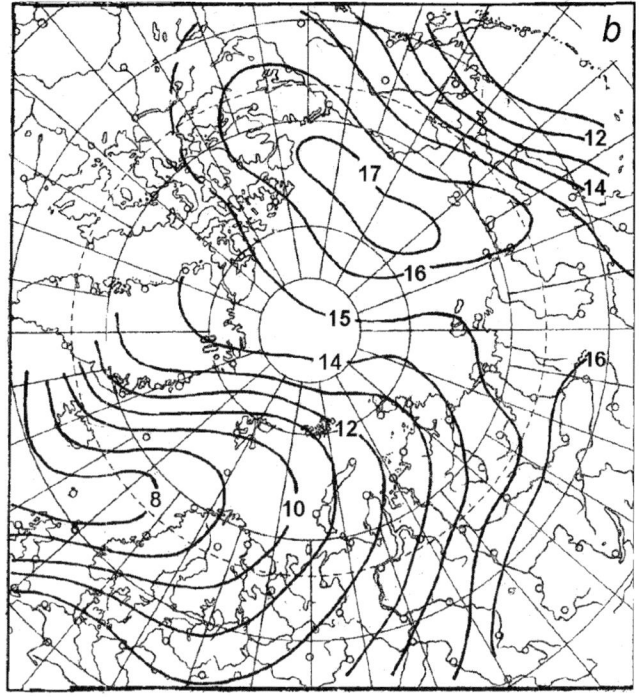

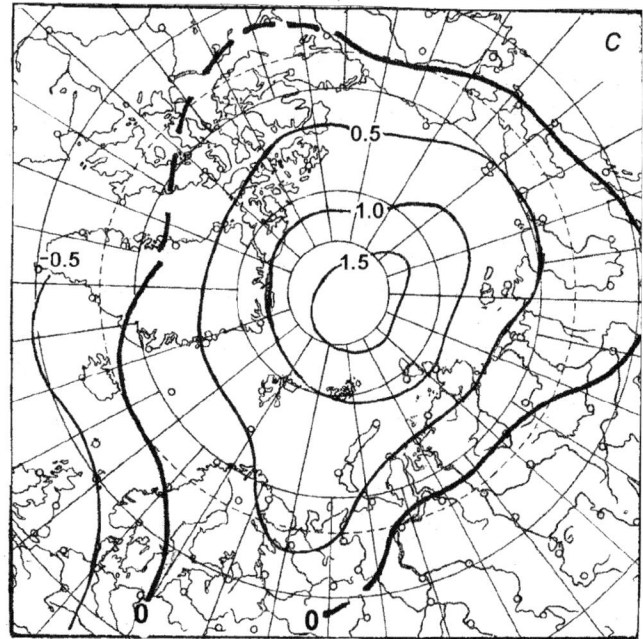

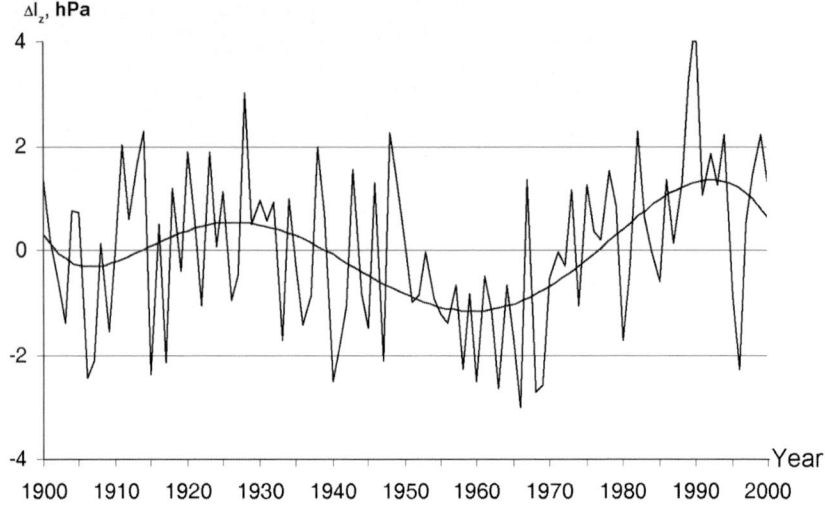

Figure 4.9. Anomalies of mean annual zonality index values in the atmosphere of temperate latitudes (40–65°N).

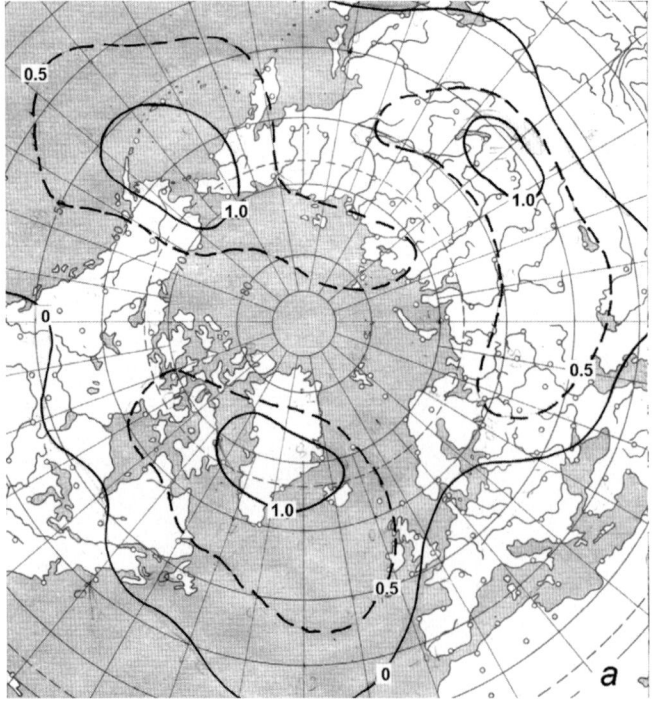

Figure 4.10. SLP anomalies (a) for the cold (1965–1975) and (b, facing page) for the warm (1990–2000) periods. (c, facing page) The difference in SLP between the warm and cold periods.

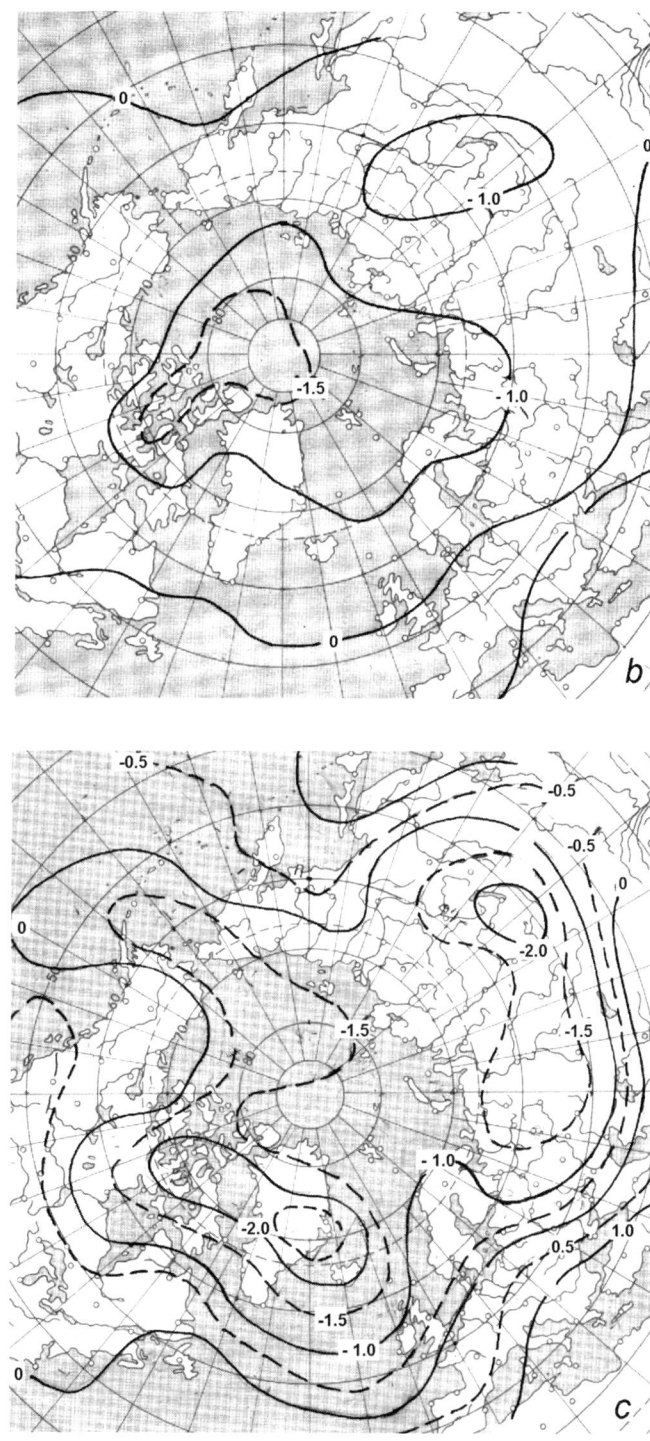

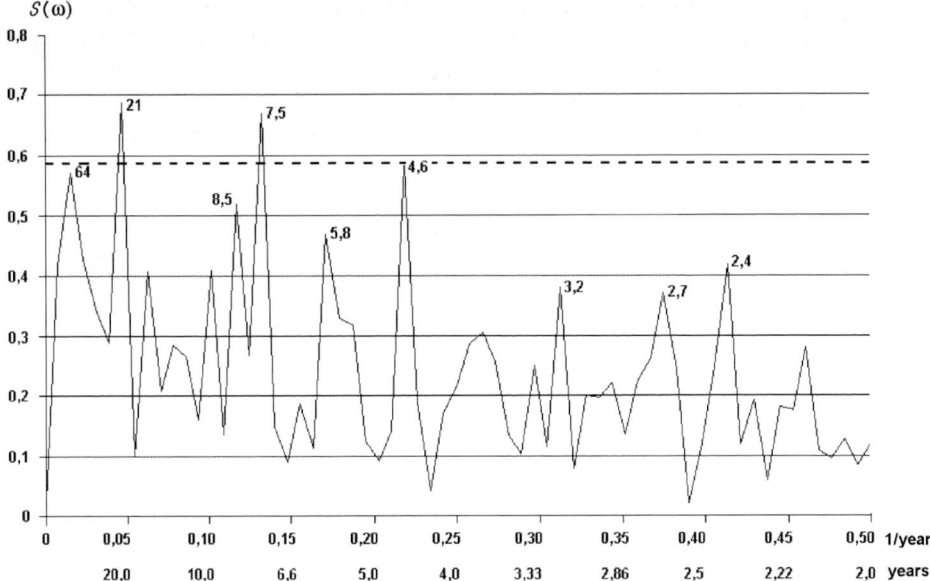

Figure 4.11. Spectrum of NAO index changes in the twentieth century. The dotted line indicates a 95% significance level.

atmosphere. Compared to air pressure field changes connected with the AO, larger-scale, 60-year cyclic fluctuations propagate to lower latitudes.

The North Atlantic Oscillation (NAO) index is another indicator of atmospheric circulation widely used in studies of climate change; it characterizes the dynamics of air mass transfer from west to east in the North Atlantic region between two powerful atmospheric pressure action centers: the Icelandic Low and the Azores High. Several variants that regulate the NAO index are considered in Smirnov *et al.* (1998); the generalized NAO index proposed by these authors is used here to characterize the change in the first main EOF expansion component of four variants of NAO indexes, averaged for December–February. The period indicated is characterized by seasonal variations in the maximum intensity of zonal transfers in the study region.

During the twentieth century, the NAO index gradually decreased from the beginning of the century to the late 1960s and then began to increase rapidly. Other significant cyclic oscillations occurred over the same time interval.

Figure 4.11 shows an NAO index spectrum calculated for the period from 1895 to 1994 by Gudkovich *et al.* (2004). The most significant oscillations occurred at frequencies corresponding to periods of about 20 and 7–8 years. Oscillations with periods of about 60 and 5–6 years are not quite as significant. A 20-year cycle in the NAO spectrum was detected earlier by Smirnov *et al.* (1998).

It is important to note that the intensity of zonal winds at North Atlantic temperate latitudes is quite closely connected with meridional air transports in Baffin Bay and the North European Basin (Alekseev *et al.*, 1998). The former is character-

ized by the East Canadian Oscillation (ECO) index and the latter by the North
European Oscillation (NEO) index. The correlation coefficient for NAO–NEO is
0.74 and for NAO–ECO, −0.69. Hence, with increasing west-to-east transfers over
the North Atlantic, the cold northerly winds over Baffin Bay and the southerly winds
bringing heat to the North European Basin intensify. On the contrary, weakening of
zonal transports results in heat advection in the northwest and cold advection northeast of the North Atlantic, which influences the Arctic climate.

Vinje (1998) reveals a close statistical relationship between the changes in the
NAO index for different time intervals (from 1 to 50 years) in the twentieth century
and April changes in ice cover area anomalies in the Nordic Seas for the same time
intervals ($R = 0.95$–0.98). The sea ice extent decrease corresponded to the increased
NAO index and vice versa.

In addition to the large-scale indexes of atmospheric circulation discussed above,
local indexes such as SLP differences over certain transects and indicators of wind
field vorticity are often used in research and operational work. Such indexes can
characterize the direction and speed of the ice drift and sea currents, heat and salt
advection, and other factors. The pressure values of the atmosphere action centers
and the locations of these centers, are often used as indices of the state and dynamics
of the atmosphere (Abramov, 1967; Maksimov and Karklin, 1969; 1970a, b).

The integral curves of corresponding anomalies are convenient indicators of
long-term changes in air transport and other hydrometeorological characteristics;
they allow us to quite clearly distinguish the time intervals during which the
anomalies predominate. Anomaly indexes (Kovalev and Yulin, 1998) made it possible to combine series with similar characteristics. We determined that negative
anomalies of mean annual air temperature and positive anomalies of total sea ice
extent of the Siberian shelf seas prevailed from the mid 1950s to the late 1980s, and
they were accompanied by weaker high-latitude zonal transports in the troposphere,
an intensified Arctic High, and other features (Gudkovich and Kovalev, 1997).

4.3   CLIMATIC CHANGES IN THE ARCTIC BASIN ICE-DRIFT PATTERN

The changes in mean SLP fields considered above also influence the corresponding
changes in the general ice-drift pattern.

In Anon. (F) and Gudkovich and Doronin (2001), the authors constructed the
mean multiyear ice-drift pattern in the Arctic Basin, which expresses the most
probable ice motion during a prescribed time interval by approximating a field of
ice-drift vectors using a two-dimensional polynomial of the form:

$$V_{x,y} = \sum_{q=0}^{t} \sum_{p=0}^{m} (a_{pq})_i x^p y^q, \qquad (4.1)$$

where x, y represent the coordinates of the initiation of drift vectors $V_{x,y}$; m is the
degree of polynomial with argument p; t is the degree of polynomial with argument q;
and a_{pq} is the polynomial coefficient.

When orthogonal components obtained from observations of ice-drift vectors of specified temporal scales (seasonal, semi-annual, and annual) are included in Equation 4.1, they form systems of conventional equations, which can be combined to create expanded matrices of normal equations. Their solution by the least-squares method allows deriving the polynomial coefficients for each component. For $m = t = 3$, there are 16 coefficients. The derived equations are used to calculate drift vectors in any regular grid within the region observed. An advantage of constructing ice-drift patterns with this method is improved spatial resolution, which is typical of calculations made by hydrodynamic models with the high reliability usually inherent in patterns based on full-scale observations.

For constructing the ice-drift patterns in the studies discussed above, we employed data from various observing platforms (drift of ship-based expeditions, "North Pole (NP)" drifting stations, ice islands, automatic stations, and radio buoys) through 1976. Most of these observations were made between 1954 and 1975, i.e., they refer to the cooling period that replaced the Arctic warming period of the 1920s–1940s (Zakharov, 1976). A new warming period began in the late 1970s and continued into the beginning of the twenty-first century. Alternation of the cooling and warming periods appears to be controlled by climatic cycles lasting about 60 years.

Using observation data on NP drifting station movement and data from "Argos" buoys, we calculated (Equation 4.1) ice-drift patterns in the Arctic Basin for the 1980–2004 warming period (Gudkovich et al., 2007). Comparing this pattern with a pattern obtained earlier illuminated the main changes that occurred in the ice-drift field at the transition of the climatic system from cooling to warming. Following the practice of Gudkovich and Doronin (2001), we investigated ice-drift vector fields for semiannual periods (October–March and April–September), using the initiating co-ordinates at 82°30′N, 180°E. The abscissa axis is directed along the 180° meridian to the south, and the ordinate axis is parallel to 270°E.

Figure 4.12a and b show the ice-drift patterns obtained by the method described above for summer and winter of the climate warming epoch that spanned the end of the twentieth and the beginning of the twenty-first century. A comparison of the patterns depicted in this figure with similar patterns published in Gudkovich and Doronin (2001), which mainly characterize the cooling period, shows no significant differences. Similar to the cooling epoch, an increase in ice-drift speed is observed during the last warming period at approaches to Fram Strait in winter, in the displacement of the transarctic flow core from Eurasia to America, and in a decrease in the area of the Beaufort anticyclonic gyre from winter to summer. The average modules of ice-drift speed for monthly and half-year time periods (3.5 and 2.5 cm/s, respectively) practically coincide.

The main differences occur in the patterns (Figure 4.12c, 4.12d) that show the differences in resulting ice-drift vectors during the warming and cooling epochs. Both summer and winter patterns clearly exhibit a cyclonic character in the vector difference fields, indicating increased cyclonicity when the cooling epoch is replaced by the warming epoch. This is expressed more strongly in summer than in winter. A decrease in Arctic atmospheric pressure during warming epochs is confirmed by the pressure charts averaged for the corresponding periods in Figure 4.10. This phenomenon

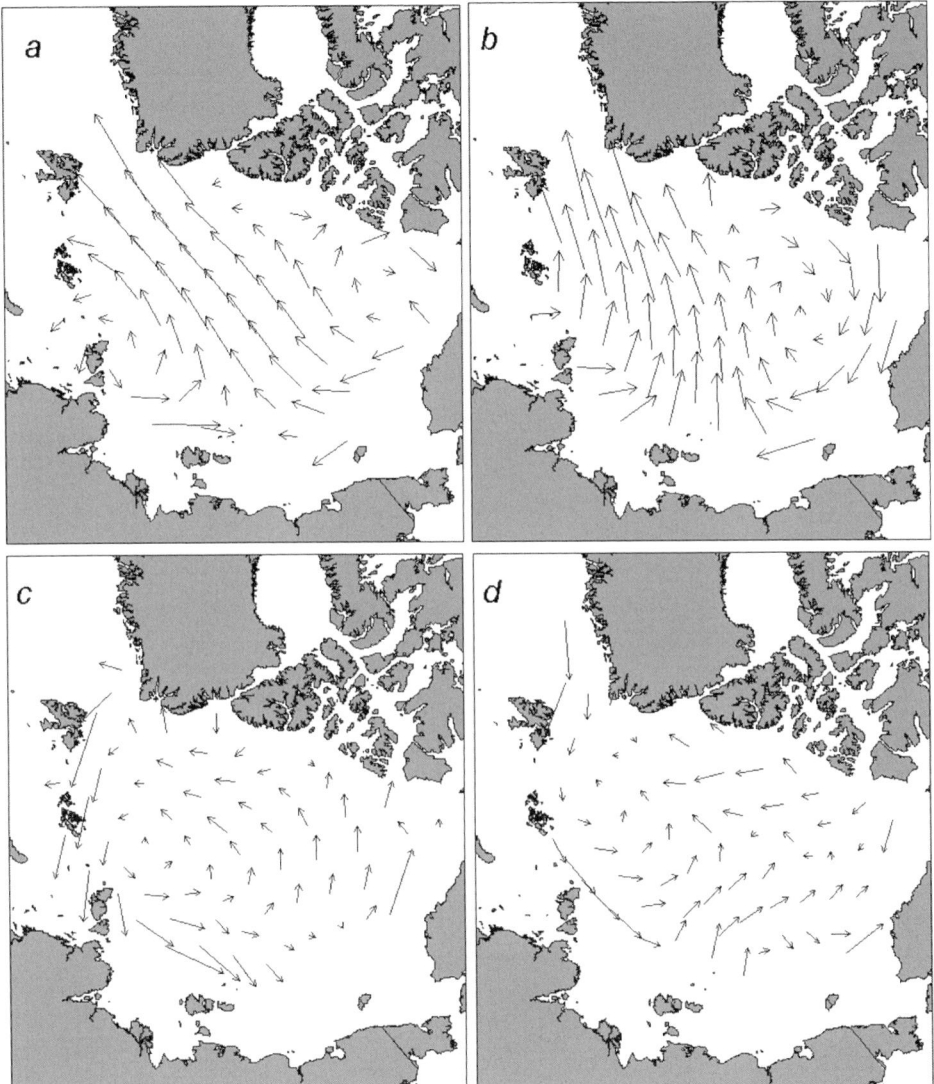

Figure 4.12. Mean resulting ice-drift pattern for summer (a) and winter (b) during the warm epoch and the difference between ice-drift vectors during the warm and cold epochs for summer (c) and winter (d).

provides a very good expression of a zonal transfer index in the atmosphere of temperate latitudes (from 40° to 65°N), similar to the Blinova (Blinova, 1943) index (Figure 4.9).

An increase in the recurrence of cyclonic pressure fields over the Arctic Basin at the transition from a cooling to a warming epoch leads to changes in ice cover deformation processes. As shown in Gudkovich and Doronin (2001), Busuyev *et al.* (1999), Porubayev (2000), Gudkovich and Klyachkin (2001), and Losev *et al.*, (2005), the cyclonic systems of the multiyear ice drift contribute to ice cover divergence. This process is most prevalent in summer, whereas in winter, especially in relatively thin ice zones, ice compacting is usually observed. Anticyclonic SLP fields have the opposite effect. This is confirmed by mathematical modeling of ice cover dynamics (Shoutilin *et al.*, 2005), showing that low-frequency changes in the state of multiyear ice in the Arctic Basin are accompanied by the processes of divergence, compacting, and ridging, which influence the corresponding changes in medium drifting ice. Divergence of multiyear ice in the Arctic Basin during warming is confirmed by experimentally detected gradual displacement of the multiyear ice boundary and its area. This is discussed in section 4.5.

Figure 4.12d shows that the differences in ice-drift vectors in the Fram Strait area in winter are directed from the Greenland Sea to the Arctic Basin. This provides further support for section 4.2 evidence that ice export from the Arctic Basin to the Greenland Sea during warming is weaker than during cooling. Section 4.4 presents additional arguments confirming a direct relationship between the ice cover area and ice export through Fram Strait.

4.4   CHANGES IN ICE EXCHANGE BETWEEN THE ARCTIC BASIN, MARGINAL SEAS, AND THE GREENLAND SEA

There is extensive sea ice exchange between the Arctic Basin and its marginal seas, which are the major sources of new ice for the Arctic Basin. The Arctic Basin serves as a reservoir for the marginal seas; it both receives large ice masses exported from the seas and supplies the seas with thicker multiyear ice. The direction and intensity of ice exchange depends to a great extent on the wind regime. However, local winds alone do not completely determine this exchange of ice. Ice export from the ice cover of marginal seas depends on sea ice conditions in the central Arctic because the sea ice originating from the marginal seas must have some ability to replace the central Arctic ice cover. Thus, the marginal seas depend to some degree on the intensity of ice export from the Arctic Basin to the Greenland and other subarctic seas. However, ice flow from the basin to the seas during onshore winds is strongly restricted by the shoreline and landfast ice, and ocean circulation also influences this ice exchange.

4.4.1   Ice export through Fram Strait

Ice export from the Arctic Basin to the Greenland Sea through Fram Strait is a significant ice balance component for both regions and an important factor in climatic changes in the Arctic. The ice exchange through Fram Strait determines most of the ice discharge of the Arctic Basin, thus influencing the thermal and dynamic processes in the basin and its marginal areas.

Ice drift observations in Fram Strait are carried out episodically. From the 1930s through the 1960s, information on ice-drift speed in the strait was obtained from observations made by a few drifting expeditions that were carried to the Greenland Sea by wind and current. Later, the amount of information increased substantially as a result of observations from drifting buoys transmitting to satellites. This information was used to develop statistical methods for calculating the ice exchange through the strait. Quite a large number of studies have been devoted to the methodology of calculating the ice quantity passing through the strait for a particular time interval (Laushkin, 1962; Gudkovich and Nikolayeva, 1963; Vowinchel, 1964; Lebedev and Uralov, 1977; Gorbunov et al., 1985; Vinje and Finnekasa, 1986; Gudkovich and Pozdnyshev, 1995; Alekseev et al., 1997). Most of the methods developed during these studies are based on the empirical dependencies of the ice-drift speed in the strait on the baric gradient value in the area of the East Greenland Current. That is why there is satisfactory correlation of data obtained by different methods on interannual and multiyear changes in the area and on the volume of ice exported through the strait. More recently, dynamic-thermodynamic sea ice models were developed, allowing the calculation of the ice transport through the strait taking into account both dynamic and thermodynamic processes (Harder et al., 1998).

The differences in various investigators estimates of the area and volume of ice exchange through Fram Strait are mainly explained by differing volumes of observational data, different approaches to determining the speed of gradient currents, and different accounts of the cross-non-uniformity of the speed of ice drift and currents and/or flow width and ice cover concentration.

Long-term data series on the speed of the mean monthly ice export to the Greenland Sea through Fram Strait were calculated by the methodology developed by Gudkovich and Nikolayeva (1963). These authors devote considerable attention to calculating the gradient ice-drift component, which plays a large role in ice drift over the entire length of the East Greenland Current (Gudkovich and Pozdnyshev, 1995). For calculating the gradient current in Fram Strait, this study employed the theory on wind currents in a baroclinic sea developed by Lineikin (1955), who derived the equations that allow calculating the speed of the wind and gradient components of the current in an infinite and deep channel, based on wind distribution across the channel. In the stationary and uniform field of tangential friction stresses directed parallel to the channel axis, the value of the desired longitudinal current speed component at the surface can be sufficiently accurately expressed by

$$U = \frac{\tau * l \sqrt{bg}}{2\pi\omega\mu},$$ (4.2)

where U is the current speed (m/s); τ is tangential wind stress; l is strait width; ω is the vertical component of the Earth's angular rotation speed; μ is the horizontal turbulent exchange coefficient; b is the vertical gradient of conventional water density; and g is gravity acceleration.

By substituting in this formula the values of $l = 45 \cdot 10^4$ m, $g = 9.8$ m/s^2, $\omega = 7 \cdot 10^{-5}$ s^{-1}, $\mu = 10^8$ kg $\cdot$ m^{-1} $\cdot$ s^{-1}, $b = 7 \cdot 10^{-5}$ m^{-1} (from data taken during the expedition onboard the diesel-electric ship Ob' in 1956), we approximately derive:

$$U = 25 * \tau$$ (4.3)

In the East Greenland Current, the sea surface is not directly influenced by the wind but rather by the ice moving relative to the water under the influence of the wind. In this case, the tangential friction is determined by the quadratic law:

$$\tau = C\rho W^2,\tag{4.4}$$

where C is the friction coefficient of the bottom ice surface; ρ is the water density; and W is the wind drift speed.

The expression of W through the atmospheric pressure gradient using linear empirical ratios allowed calculation of a family of parabolas corresponding to different values of the friction coefficient C. The results were compared with the gradient drift speed from NP-1 station records (December 1937) and the icebreaking ship $G.\ Sedov$ (December 1939–January 1940). To average the speeds, the locations of these expeditions were taken into account relative to the outflow cores, which were usually situated near the continental slope. The results of dynamic processing of hydrographic observations in the strait area carried out by the $Sever$-7 and Ob' expeditions (1955 and 1956, respectively) were also used. The best match between the observed data and Equation (4.4) was for $C = 0.004$. The calculated water velocity was smaller than observed by approximately 0.025 m/s. This value (0.025 m/s) characterizes the water current which is determined predominantly by river runoff and inflow of Pacific water to the Arctic Ocean via Bering Strait (due to sea level difference between the Pacific and Atlantic Oceans) (Proshutinsky, 1993).

As a result, the empirical expression for calculating the average speed of the export current in Fram Strait (in m/s) for each month had the form:

$$U = 270 \cdot G^2 + 0.025\tag{4.5}$$

where G is the average baric gradient (hPa/km) calculated for two sections: from Cape North-East (Greenland) to Amsterdam Island (Spitsbergen), and from Clavering Island ($73°30'$N, $21°30'$W) to a point at the intersection of the parallel $70°$N with the Greenwich meridian. The value of G was averaged for three preceding months, including the month for which the current speed was calculated. This accounted for the assumed time for establishment of baroclinic currents.

The speed of the wind component of the ice exchange through the strait was calculated using mean monthly baric gradients in Fram Strait, perpendicular to the strait axis, using the known values of isobaric coefficients (Gudkovich and Nikolayeva, 1963). The speed values derived were added to the corresponding gradient drift speed values. The average width of the ice flow in the strait was assumed to be 340 km.

The use of this methodology made it possible to calculate a data series for the area of ice exported through Fram Strait for each month of the twentieth century. An archive of mean monthly atmospheric pressure charts for 1900–2000, available at the AARI, was used to calculate baric gradients.

A 100-year analysis of mean monthly ice exchange between the Arctic Basin and the Greenland Sea showed that the most reliable data became available in the late 1920s to the early 1930s, when atmospheric pressure information in the area adjoining Fram Strait became more or less accurate. Due to a gap in observations during

World War II, from 1941 to 1945, analysis of the calculated data was performed for a number of years from 1946/1947 to 2002/2003. Based on these data, the mean annual area of ice exported to the Greenland Sea comprises 650,000 km^2. Based on satellite passive microwave data, Kwok, Cunningham and Pang (2004) estimated the mean annual ice export for 1978–2002 as 866,000 km^2. Both values are much lower than the estimates published by Gordienko and Karelin (1945) and Vinje (1992): 1.04 and 1.08 million km^2, respectively.

A possible cause of overestimation of the ice exchange values from observations of the drift of radio buoys near Fram Strait might be a rapid increase in drift speed moving southward, which allows incorporating only the resulting vectors for comparatively short time intervals (up to a week). Note that Gudkovich and Doronin (2001) found that the average ice-drift speed increases with a decreasing period of averaging. On the other hand, due to a large cross non-uniformity of ice export speed, most observations characterize conditions near the current core, where drift speeds are much higher than they are on average along the transect.

An analysis of the data using the methodology described above shows that the intensity of ice exchange through Fram Strait changes throughout the year, increasing in winter and decreasing in summer (Figure 4.13). On average, two-thirds of the annual ice export occurs from November to April and only one-third from May to October. There are significant interannual fluctuations in the ice exchange area. A spectral analysis reveals cycles lasting 8–10 years and about 2–3 years.

Vinje and Finnekasa (1986) calculated the average annual speed of ice export through Fram Strait using Arctic buoy drift data for 1976–1984 obtained during the ICEX (Ice Experiment), AOBP (Arctic Ocean Buoy Program), and MIZEX (Marginal Ice Zone Experiment) programs. They showed that the ice transport ranges from $125 \cdot 10^3$ to $173 \cdot 10^3$ m^3/s (with a mean of $153 \cdot 10^3$ m^3/s). These authors derived the regression equation relating the ice export speed for weekly time intervals (Q, m^3/s) to the atmospheric pressure difference (ΔP, hPa) between 81°N, 15°W and 73°N, 5°E:

$$Q = (90.5 + 9.6\Delta P) \cdot 10^3 \tag{4.6}$$

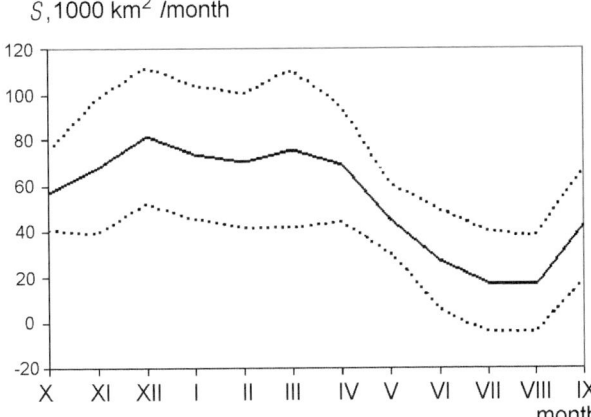

Figure 4.13. Average annual variations of the area of mean monthly ice export from the Arctic Basin to the Greenland Sea through Fram Strait.

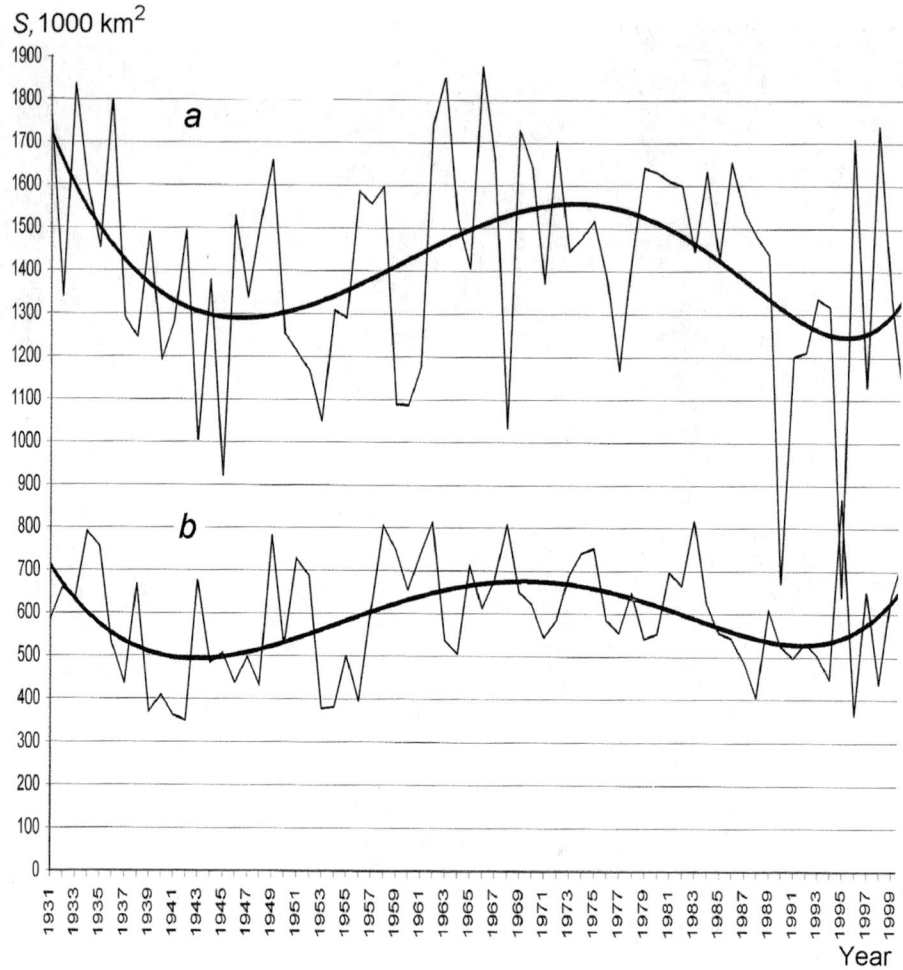

Figure 4.14. (a) Interannual fluctuations of the total ice area of the Siberian shelf seas in August, and (b) areas of ice exported from the Arctic Basin through Fram Strait. The values of the bold curves are smoothed by a polynomial to the power of 6.

The correlation coefficient of this relationship for winter is 0.95. However, the values of Q strongly depend on the estimate of average ice thickness in the strait, which varies from 2.66 to 4.06 m in the calculations of different authors.

Based on the methodology of Gudkovich and Nikolayeva (1963), with a correlation coefficient of 0.91, Alekseev *et al.* (1997) derived a regression equation relating the value of Q calculated by Equation 4.6 with the total ice exchange area (X, km^2) for the winter:

$$Q = (96.4 + 0.0013 \cdot X) \cdot 10^3. \tag{4.7}$$

Figure 4.14b shows long-period changes in the total area of ice export through Fram Strait from October of one year to August of the next year for 1931–2000. An

Table 4.4. Correlation coefficients between the long-period fluctuations of the area of ice exported through Fram Strait (October–August) and total ice area of the Arctic Seas Asian shelf in August for the period 1931–2000 at different time lags.

Time lag (years)	0	1	2	3	4	5	6	7	8	9	10	11	12
Correlation coefficients	0.43	0.56	0.67	0.75	0.80	0.81	0.80	0.77	0.72	0.66	0.58	0.49	0.39

approximation of data by a polynomial to the power of 6 (bold curve) indicates the cyclic character of these changes, with the cycle lasting about 60 years. Figure 4.14a shows that the fluctuations of total sea ice extent of the Arctic Seas of the Siberian shelf (from the Kara to the Chukchi Seas) have a similar character.

It is remarkable that increased ice export through Fram Strait is accompanied by increased sea ice extent in the Arctic Seas, contrary to the opinions of those who assume that ice export to the Greenland Sea increases during climate warming, accompanied by a decrease in sea ice extent in the Arctic Seas (Rigor *et al.*, 2002; Makshtas *et al.*, 2002; Hassol, 2004).

As shown in Figure 4.14, ice export fluctuations slightly precede corresponding sea ice extent changes in the Arctic Seas. The cross-correlation function between the smoothed values of ice export and total sea ice extent exhibits the highest correlation coefficients at time lags (sea ice extent after export) of 4, 5, and 6 years (Table 4.4). Following decreased ice export through Fram Strait in the early 1990s, a tendency for its increase was observed. Based on the time lags shown in Table 4.4, a transition to the phase of increased sea ice extent in the Arctic Seas would be expected at the beginning of the twenty-first century, as confirmed by Figure 4.14a.

So, fluctuations of ice export through Fram Strait occur approximately 4–6 years ahead of total sea ice extent fluctuations in the Arctic Seas. However, ice export to the Greenland Sea has a different influence on the sea ice extent of various seas. In this regard, it is of interest to compare the mean annual area of ice export from the Arctic Basin with the difference of sea ice extent in the Severozemelsky region (northeastern Kara Sea and western Laptev Sea) and the Wrangelevsky region (eastern East Siberian Sea and southwestern Chukchi Seas) in August. This difference characterizes the general distribution of the ice cover along the Northern Sea Route. It turns out that the correlation coefficient for this synchronous relationship is +0.40 (or 0.23 at $P = 95\%$ confidence level). With a 7-year shift (the difference in sea ice extent after ice export), the correlation coefficient increases to 0.62. So, 7 years after an increase in ice export from the Arctic Basin, the sea ice extent of the Severozemelsky region increases and that of the Wrangelevsky region decreases, and vice versa. This repeated pattern confirms the results mentioned above, in particular the opposing ice conditions in the western and eastern Arctic Seas.

A phenomenon termed "speed leveling along the general drift of the export flow" by Volkov and Gudkovich (1967) causes ice export through Fram Strait to affect ice cover dynamics in the Arctic Basin and its marginal seas. It is also possible that the inflow of Pacific Ocean water through Bering Strait increases when there is an increase in water and ice export through Fram Strait and decreases at its attenuation (Gudkovich, 1961).

The average drift and current speed in Fram Strait for the preceding year influences the ice exchange between the Arctic Basin and the Laptev, East Siberian, and Chukchi Seas in winter (October–March). The increased ice export to the Greenland Sea contributes to the increased ice export from these seas to the Arctic Basin, and its decrease results in the opposite effect (Gudkovich and Nikolayeva, 1963).

A 20-year observation series by Gudkovich and Kovalev (1967) shows that the latitude of the multiyear ice boundary in spring and the ice edge in the subsequent autumn in the northern Chukchi Sea (at 185–190°E) depends on the anomalies of ice export through Fram Strait. The correlation coefficient achieves its maximum value (0.86) if the ice export value is averaged for two preceding years. In the same study, a correlation of ice export to the Greenland Sea with the area of anticyclonic water circulation in the sub-Pacific Ocean sector of the Arctic Basin and the location of the core of the Transarctic current was revealed: upon increased ice export, the circulation expands and its core moves to the west; decreased export has the opposite effect. A cycle of 8–10 years was noted for these changes.

The influence of ice flow from the Arctic Basin to the Greenland Sea on the sea ice extent of the East Greenland ice belt was included in ice balance calculations by Lebedev and Uralov (1977). They concluded that in addition to ice exchange, the ice area in this region depends on the thermal processes (ice formation and melting) taking place. Similar results were obtained by Moritz (1988).

4.4.2   Ice exchange between the Arctic Seas and the Arctic Basin

Ice exchanges between the marginal seas and the Arctic Basin, along with thermodynamic processes, influence the ice cover structure in these seas and hence their sea ice extent (Gudkovich and Nikolayeva, 1963; Gudkovich *et al.*, 1972; Gudkovich and Doronin, 2001).

Unfortunately, no direct, sufficiently long-term ice-drift observations are available near the boundaries between the marginal seas and the Arctic Basin, and calculation methods must be used for estimating the ice exchange (ice area or volume) and its variability in time. In the "export" seas, which contribute ice to the Arctic Basin during much of the year, the simplest calculation method is based on "isobaric drift" ratios proposed by N.N. Zubov (1944):

$$w = k\frac{\partial P}{\partial x}, \qquad\qquad (4.8)$$

where w is the projection of the ice-drift speed to axis y; k is the isobaric coefficient; and $\dfrac{\partial P}{\partial x}$ is the projection of the atmospheric pressure gradient on axis x.

If axis x is directed along the "entry" section with a length l, approximately coinciding with the northern boundary of the sea, then integrating Equation 4.8 along this axis from the western to the eastern sea boundary results in:

$$S = \int_0^l w\, dx = k \int_0^l \frac{dP}{dx}\, dx = k(P_l - P_0) = k\,\Delta P. \tag{4.9}$$

Here, S is the area of the ice cover passing through section l at unit time. This area is determined by the dimension of coefficient k and the corresponding scale of averaging of the baric chart. When using the mean monthly charts of atmospheric pressure at sea level, the isobaric coefficient dimension is $km^2/hPa \cdot month$. These values were derived from observations of the total ice drift in the Arctic Basin, increased by 25%, according to the empirical ratio of corresponding mean annual values (Gudkovich and Nikolayeva, 1963).

Equation 4.9 indicates that the resulting ice exchange area does not clearly depend on section length and is proportional to the atmospheric pressure difference at its ends. It is assumed that the ice cover (regardless of ice concentration) does not disappear along the entire section length. Thus, a correct estimate of the ice exchange volume requires information on ice concentration and thickness and their changes in time.

Estimates of the area of ice exchange between the Barents, Kara, and Laptev Seas and the Arctic Basin for 1937 to 2003 were based on monthly differences in atmospheric pressure between Spitsbergen and Franz-Josef Land, between Franz-Josef Land and Severnaya Zemlya, and between Cape Arktichesky (Severnaya Zemlya) and Kotel'ny Island (Novosibirskie islands) with account for the monthly values of isobaric coefficients published in Gudkovich and Nikolayeva (1963). Using these data, seasonal changes in the corresponding mean multiyear values shown in Figure 4.15a, b, c were determined: in summer (mainly from May to August), ice is exported to these seas from the Arctic Basin, and in winter (mainly from September to March), ice is exported from these seas to the Arctic Basin. These data suggest that more than 40,000 km^2 (with a standard deviation of about 70,000 km^2) is exported on average from the Barents Sea to the north, about 120,000 km^2 (standard deviation of about 90,000 km^2) from the Kara Sea, and more than 290,000 km^2 (standard deviation of more than 90,000 km^2) from the Laptev Sea in winter. The ice export from the Arctic Basin in summer comprises on average about 25,000 km^2 for the first two seas and about 10,000 km^2 for the Laptev Sea (standard deviations are about 50,000–70,000, and more than 115,000 km^2, respectively). There is significant ice exchange between the Barents and Kara Seas. For much of the year (from August to June), ice is exported from the Kara Sea to the Barents Sea. The area of this ice is comparable to the ice export from the Kara Sea to the Arctic Basin for the winter period. These estimates do not account, in explicit form, for the influence of the gradient currents, which may influence the values given above.

According to Gudkovich and Nikolayeva (1963), in a year that westerly and southwesterly winds increase over the eastern Barents Sea during October–December, the setup they create in the Kara Sea increases ice export from this sea toward the north. Dominant easterly and northeasterly winds produce the opposite result. This study also shows that ice export from the eastern East Siberian Sea and

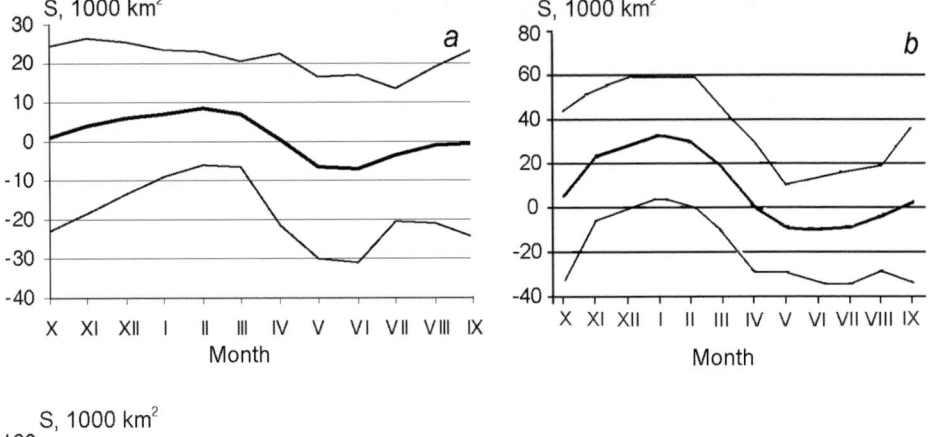

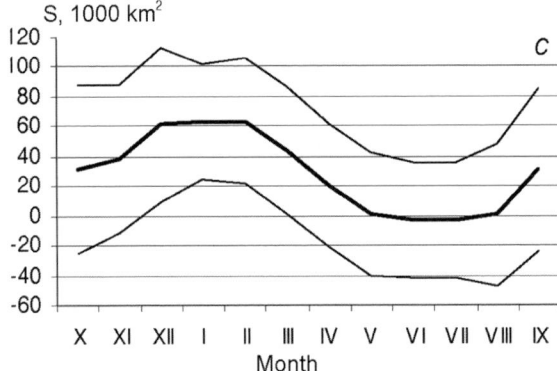

Figure 4.15. Mean multiyear values of seasonal changes in the calculated ice exchange of the Barents (a), Kara (b) and Laptev (c) Seas with the Arctic Basin (thin lines characterize data that were increased or decreased by standard deviation values).

the southwestern Chukchi Sea during the period considered is strongly influenced by wind field vorticity in the vicinity of Wrangel Island. Anticyclonic vorticity increases the ice export, and cyclonic vorticity results in additional ice flow from the north.

To check the reliability of the above calculations, the boundary of ice with total concentration 7–10 tenths in late September in the Laptev Sea was compared with the boundary of old (second- and multiyear) ice dominance (partial concentration 5 tenths and greater) in March of the following year. The difference in the latitude of these boundaries at meridians spaced at 5° of longitude was assumed to be the value of ice motion along the corresponding meridians for the six winter months (October March). The ice exchange area of the seas with the Arctic Basin (S) for the designated period was determined by:

$$S = 12350 \cdot \Delta\varphi_{mean} \cdot \Delta\lambda \cdot \cos \varphi_{mean} \qquad (4.10)$$

where $\Delta\varphi_{mean}$ is the average difference in latitude between the aforementioned ice boundaries at meridians (in degrees); $\Delta\lambda$ is the difference in longitude between the meridians of the eastern and western boundaries of the sea (in degrees); φ_{mean} is the latitude of both ice boundaries averaged by meridian, and a coefficient of 12350 converts the degrees of latitude to km^2.

The average annual ice exchange of the Laptev Sea with the Arctic Basin for the winters from 1954 to 2002 calculated using Equation 4.10 is 294,000 km^2 or 55% of the sea area, which essentially coincides with the value presented above that was calculated from the atmospheric pressure difference. Calculations of the ice exchange area for other seasons are not expected to result in significant differences. Adding 55,000 km^2 of the calculated average ice export from April to September to this winter ice exchange value results in an estimate of 349,000 km^2, assuming significant interannual variability does not cause this estimation to differ strongly from average ice export from this sea to the Arctic Basin (309,000 km^2) for the years 1936 to 1995, as calculated by means of a semi-empirical model described by Alexandrov et al., (2000). These authors modeled ice drift using a large-scale dynamic-thermodynamic model to determine the relationship of ice export from the Laptev Sea to the north and east with alternating cyclonic (CR) and anticyclonic (AR) circulation regimes. The authors concluded that during an AR, ice export increases to the north and decreases to the east. A CR has the opposite effect, with weaker ice export to the north and stronger export to the east. A similar phenomenon was noted during increased cyclonic activity over the Arctic Basin during the first twentieth century Arctic warming when the icebreaking vessel *G. Sedov* drifted during the winter of 1937–1938 to the east and onto the shelf north of the New Siberian Islands. An east current that appeared at the time was later called the "*G. Sedov* current." This current was probably absent during the Fram's drift, when there was a cold period similar to that of the 1960s–1970s.

Estimating ice exchange between the Arctic Basin and the East Siberian and the Chukchi Seas is more complex. During the winter period, onshore winds are often observed here, which should result in the export of a large amount of ice from the Arctic Basin to these seas. The intensity of this process increases from west to east. However, our analysis of ice area change in the eastern part of the East Siberian Sea and in the Chukchi Sea shows that throughout the winter, ice export from these seas to the Arctic Basin dominates (Gudkovich and Nikolayeva, 1963).

This conclusion is also confirmed by other data from the present study: The area of ice exported to the Arctic Basin from the East Siberian Sea during the winter period comprises only 38,000 km^2; however, determining its value from the change in location of the close ice boundary (7–10 tenths) in late September, and of the prevailing old ice boundary in March, (Figure 4.16) yields more than 355,000 km^2. About 100,000 km^2 of this quantity is located to the north of the New Siberian Islands, and should mainly comprise additional ice exported from the Laptev Sea. However, in this case, ice export from the East Siberian Sea also contributes substantially to the movement of ice from the shelf seas to the Arctic Basin. It should be noted that all data on the position of the ice edge were obtained by means of processing AARI routine 10-day periodicity ice charts in the WMO SIGRID format for 1954–1992, i.e., for the period with the smallest number of gaps in the historical dataset (see Mahoney et al. (2008) for a full description of the dataset).

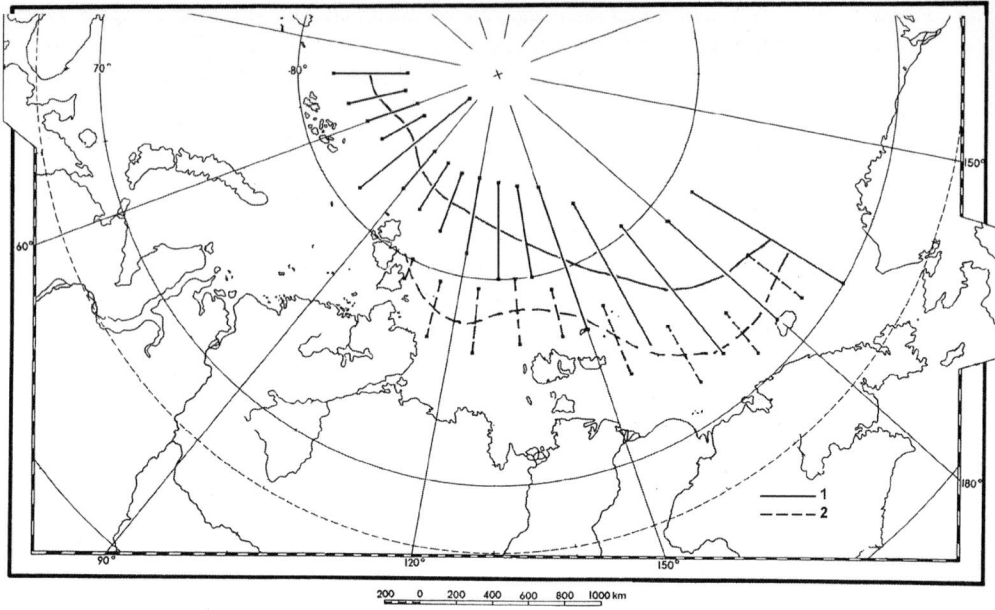

Figure 4.16. Average 1954–1991 boundaries of prevailing old ice in March (1) and close residual ice in late September of the preceding year (2). Line segments at meridians characterize corresponding standard deviations.

In the Chukchi Sea, onshore winds prevail on average much of the year (except for June–July). The winter ice flow from the north calculated by the atmospheric pressure difference along the Wrangel Island to Cape Lisborne transect comprises more than 300,000 km^2 on average, whereas the displacement of ice boundaries indicates the dominance of ice export to the Arctic Basin (on average, 14,000 km^2). These differences are attributed, as noted above, to the influence of the Pacific Ocean current, whose speed increases as it approaches the Bering Strait, and by resistance of the internal ice cover to compression processes.

It should also be noted that dynamic divergence of the ice cover near the northern sea boundary, which is often observed during prevailing cyclonic baric fields in the summer, can slightly influence the location of the boundary of prevailing old ice at the end of winter and hence cause overestimation of ice export to the north obtained when using the methodology described.

While the factors discussed above introduce large errors in calculations of ice exchange between the East Siberian Sea and the Arctic Basin in winter, spring–summer (April–September) calculations are significantly more reliable, because offshore winds tend to dominate during this period. The total calculated area of ice exported to the Arctic Basin for in the spring-summer season comprises 105,000 km^2, on average, with a standard deviation of 127,000 km^2.

It is impossible to estimate ice exchange between the Chukchi Sea and the Arctic Basin in summer; an intensified Pacific Ocean current in the Bering Strait at this time and heat advection in the atmosphere usually cause a significant part of the ice cover to melt within the sea. In some years, when an extended baric depression is established in the adjoining areas of the Arctic Basin, rapid movement of a large amount of ice from the Arctic Basin to the eastern East Siberian Sea and the Chukchi Sea can occur.

The estimates above show that, on average, about 1 million km^2 of the ice cover is transported annually from the Arctic Seas to the Arctic Basin, which is comparable to current estimates of the area of ice exported annually from the Arctic Basin to the Greenland Sea. (e.g., Koesner, 1973; Mironov and Uralov, 1991; Vinje, 1986). Given a typical ice thickness value, we can estimate the volume of ice exported to the Arctic Basin during a winter to be approximately 1500–2000 km^3. This value is about half as large as the available estimates of ice export to the Greenland Sea in winter (Vinje and Finnekasa, 1986; Alekseev et al., 1997), which can be accounted for by ice growth, ice ridging, and other processes that occur during transport of the ice to Fram Strait.

As the thickness of the ice cover involved in ice exchange constantly changes, the best approach for estimating corresponding ice volume should be based on the dynamic-thermodynamic models of ice cover evolution. This requires performing model calculations for long time intervals (years), during which the resulting drift speed becomes comparable to the systematic error of calculation (Gudkovich and Doronin, 2001). The models should be improved, especially to account for gradient currents and rheological properties of the ice cover.

As shown in Frolov et al. (2005), Gudkovich et al. (1972), and Gudkovich and Doronin (2001), the ice exchange of the marginal seas with the Arctic Basin in winter influences the formation of macro-scale ice structures, expressed in ice zones of different age, and hence of different thickness. The ice melting rate and the disappearance of ice in the seas the following summer depends on the latter. Due to various hydrometeorological conditions typical of different seas, there are differences in the thicknesses of ice zones of the same age. The areas of the corresponding zones determined by the ice exchange intensity during a particular period also differ significantly. As a result, non-deformed ice formed during the autumn-winter period in the Barents Sea typically is not thicker than 100 cm by the beginning of spring melting. Average ice thickness in the Kara Sea is 130–170 cm, in the western Laptev Sea 180–190 cm, in the eastern Laptev Sea and the western East Siberian Sea 200–215 cm, and in the Chukchi Sea 150–170 cm.

When spring–summer melting occurs at average intensity, the anomalies of the area of ice exchange and ice growth in winter almost completely determine the sea ice extent of the Barents Sea the next summer. In the other Arctic Seas the influence of ice exchange with the Arctic Basin on the subsequent disappearance of ice is shifted to the end of winter as ice of earlier formation will not have time to melt during a short Arctic summer. The closest connection between sea ice extent and ice exchange is observed during April–June (or May–July), when favorable (or unfavorable) anomalies in ice exchange with the Arctic Basin are accompanied by corresponding air temperature anomalies that initiate melting and seasonal changes in ice cover reflectivity (albedo).

Examining mean multiyear data on the ice exchange of the seas with the Arctic Basin in the summer may lead to an incorrect conclusion that ice exchange during this time interval plays a small role in the ice balance of the seas: the area of ice brought to

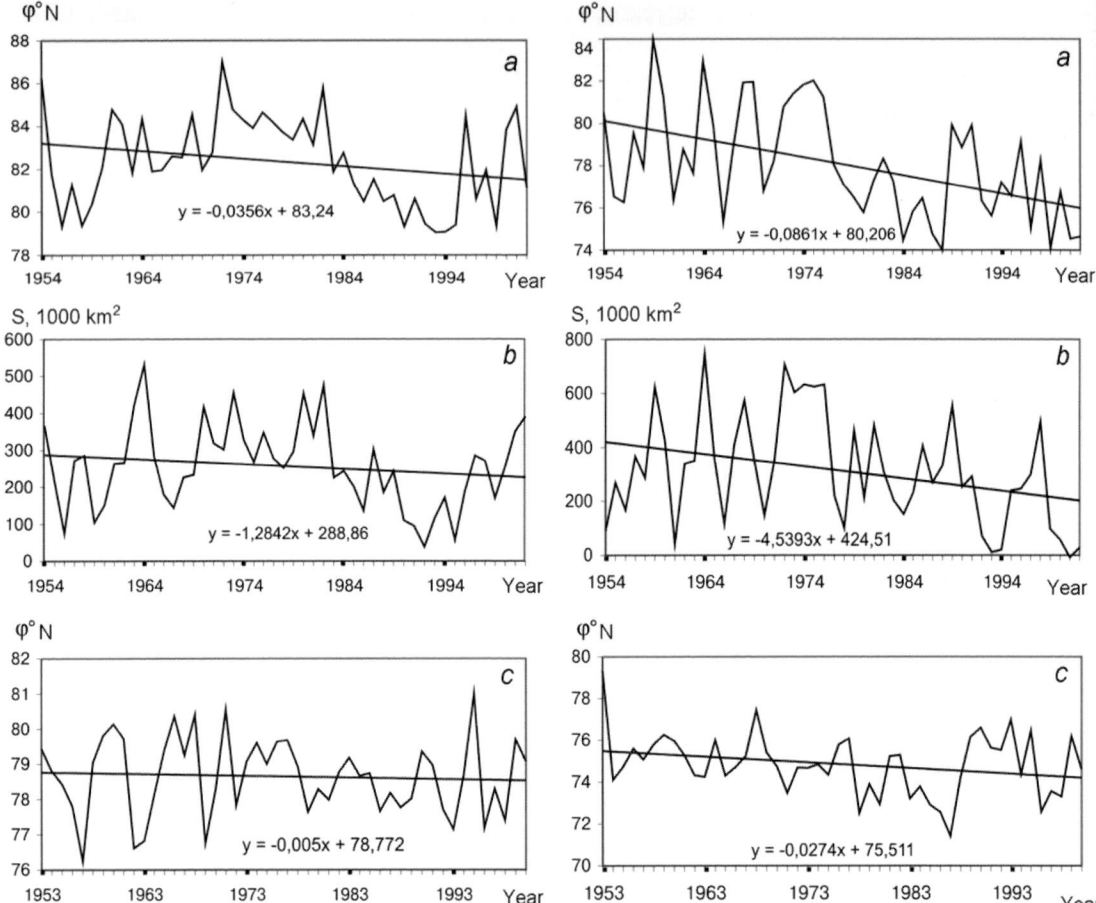

the sea or exported to the Arctic Basin from June to September comprises on average only 3%–8% of the area of the seas. However, the interannual variability of these values is quite significant. The amplitude of ice exchange fluctuations compared to the area of the sea for June–July is 22% for the Kara Sea, 42% for the Laptev Sea, and 24% for the East Siberian Sea; for the June-to-August period, the fluctuations in the three seas are 23%, 48%, and 40%, respectively, and for the June-to-September period, they are 37%, 70%, and 64%, respectively (Gudkovich and Doronin, 2001).

So, the direct role of summer ice exchange with the Arctic Basin in the ice balance of the seas during anomalous years is significant, especially for the Laptev Sea and the East Siberian Sea. In these seas, its absolute value for June–September in 30% of cases exceeds 20% of the area of the seas, while in the Kara Sea, this occurs in only 15% of cases. This component plays an even smaller role in the summer ice balance of the Barents Sea.

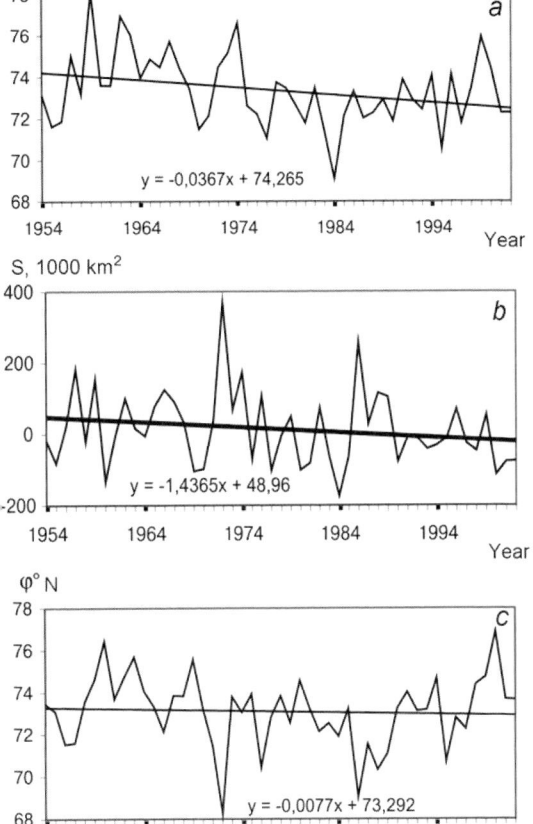

Figure 4.17. (a) Changes in the average latitude of prevailing old ice boundaries at the end of winter. (b) Ice export to the Arctic Basin for the winter period. (c) Average latitude of the boundaries of residual ice at the end of summer (in the preceding year) in the Laptev (on the left), the East Siberian (at center), and the Chukchi (on the right) Seas.

4.5 LONG-TERM CHANGES IN MULTIYEAR ICE EXTENT IN THE ARCTIC BASIN

The characteristics of multiyear ice, which include its area, thickness, concentration, and location, are all important indicators of climatic change in the Arctic Ocean. In addition, successful high-latitude passage of transport vessels and icebreakers as well as the survival of drifting research stations depend on this ice. Thus, there is clear value in studying the processes of multiyear ice formation and the variability of its boundaries. Its observations were drawn from AARI routine 10-day ice charts and provide a basis for analyzing the long-term variability of the multiyear ice boundary to the north of the Arctic Seas on the Siberian shelf for half a century.

Figure 4.17 shows changes in the locations of boundaries of prevailing old ice in February–March, close residual ice during the third 10-day period of September in the preceding year at meridians of the Laptev, East Siberian, and Chukchi Seas, and

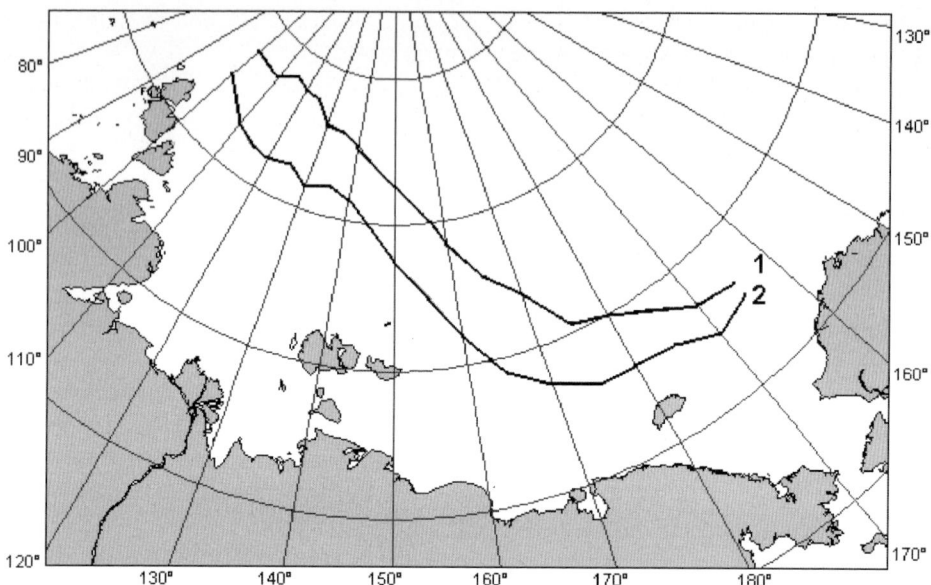

Figure 4.18. Average location of the old ice boundary in the eastern Arctic Seas for the periods 1960–1979 (1) and 1980–2000 (2).

the winter ice exchange of these seas with the Arctic Basin calculated using Equation 4.10 for 1954–2002. All curves in this figure include negative linear trends. Most significant are the trends in the location of old ice (their contributions to variability range from 6% to 22%); the least significant trends are the changes in the boundaries of residual ice (where the contribution to variability is 0.4% to 7%). The linear trends of the East Siberian Sea make the largest contribution to variability.

A study of the average location of the winter boundary of prevailing multiyear ice (with a concentration of more than 5 tenths) during the 20-year period from 1960 to 1979 and the subsequent 20-year period from 1980 to 2000 showed consistent southward displacement of the boundary (Figure 4.18). On average, it moved over 300 km southward over the 40-year period. This displacement usually took place when there was a small negative trend in the region's sea ice extent (see Figure 2.3). The cause of the southward displacement of close ice in this region is a noticeable weakening of the Arctic High (with an increased number of cyclones) toward the end of the twentieth century, accompanied by a diverging ice cover (see also section 4.3), and the aforementioned decrease in ice export from the Arctic Basin to the Greenland Sea during the warm periods as compared with the cold periods.

This southward displacement of the prevailing multiyear ice boundary is confirmed by Asmus *et al.* (2005), who noted a small positive trend, a 5% increase in the area of multiyear ice in the Arctic Basin between 40°E and 105°E, based on the analysis of ice charts constructed from radar and microwave satellite images for the period 1983–2005. Hence, there was an expansion of the multiyear ice zone during the period of climate warming not only in the eastern sector of the Eurasian part of the Arctic Basin, but also in its western sector.

The linear trends shown in Figure 4.17 are accompanied by cyclic fluctuations of varying duration. The time interval examined here includes an epoch of elevated sea ice extent from the late 1960s to the early 1980s and partly covers the epochs of decreased sea ice extent in the late 1950s and in the 1980s–1990s. The ice exchange of the seas with the Arctic Basin clearly depends upon the sea ice extent of the region and ice export to the Greenland Sea through Fram Strait. For estimating this relationship, the average values of ice exchange of the seas with the Arctic Basin were determined for the years from 1965 to 1985 ("cold" epoch) and for the combined time intervals of 1954–1964 and 1986–1995 ("warm" epochs). A comparison of these values showed average ice export from the Laptev Sea in the "cold" epoch was 42% higher than in the "warm" epoch. There was a similar, but smaller, excess of ice export from the East Siberian Sea of 25%. On the contrary, the ice export from the Chukchi Sea in the "warm" epoch was higher than in the "cold" epoch by almost 60%; This is caused by increased cyclonicity in the baric field over the Arctic Basin during the warm epoch, which leads to increased recurrence of southerly winds to the north of the Chukchi Sea.

These results are based on the reasonable assumption that the location of the old ice boundary to the north of the marginal Arctic Seas at the end of winter depends both on the intensity of ice export from these seas and on the location of the boundary of prevailing residual ice at the end of the preceding summer. However, each of these factors plays a different role in different seas, as indicated by the correlation coefficients presented in Table 4.5.

As Table 4.5 shows, the Laptev Sea exhibits the closest relationship between the location of the boundary of prevailing old ice and the location of close residual ice during the preceding summer; this relationship diminishes noticeably with increasing distance eastward from the Laptev Sea. The relationship with winter ice export to the Arctic Basin is slightly greater in the East Siberian Sea. The methodology described in Anon. (H) was used to estimate the role of both factors in the multiple regression equations connecting the boundaries of old and residual ice and the ice exchange of the seas with the Arctic Basin. Table 4.6 shows the results of this assessment.

Table 4.5. Correlation coefficients of relationships between the boundaries of (X) prevailing old ice at the end of winter, (Y) the value of the marginal seas' ice exchange with the Arctic Basin, and (Z) the boundary of close ice at the end of the preceding summer in the Laptev, East Siberian, and Chukchi Seas

Connection	Laptev Sea	East-Siberian Sea	Chukchi Sea
X–Z	0.94	0.45	0.36
X–Y	0.63	0.73	0.58
Y–Z	0.33	–0.27	–0.55

Table 4.6. Role of variability in the location of residual ice (Z) and ice exchange with the Arctic Basin (Y) in the formation of the old ice boundary at meridians of the Laptev, East Siberian, and Chukchi Seas at the end of winter

Sea	Z	Y
Laptev	77	23
East Siberian	33	67
Chukchi	59	41

The cause of the regional differences is probably the difference in the prevailing geographical location of residual ice (its remoteness from the shore, shelf boundaries, landfast ice, Bering Strait, etc.), because the speed and stability of currents and action of internal forces in the ice cover depend on it. The significance of such factors is indicated by the correlation coefficients shown in the bottom line of Table 4.5. As this table shows, the further south the location of residual ice in the Chukchi Sea, the more intensive is its export from the sea. This pattern is less evident in the East Siberian Sea and changes sign in the Laptev Sea, where the ice export slightly increases with northward displacement of the residual ice boundary.

Regarding the influence of the location of the old ice boundary at the end of winter on the sea ice extent the following summer, the correlation analysis shows that the corresponding correlation coefficients (0.20–0.25) are statistically insignificant, although the negative sign of the two eastern seas' coefficients indicates a possible weak dependence: the more northern the old ice boundary, the smaller the sea ice extent.

4.6   LONG-TERM CHANGES IN SOME WATER MASS CHARACTERISTICS OF THE ARCTIC OCEAN

The current state of Arctic Ocean water and its long-term variability are important factors because they reflect changes in global climate and influence these changes. Salinity is of special significance because water density and hence water dynamics (currents, convection) depend on it at high latitudes. The vertical distribution of water density influences heat exchange between the ocean and the atmosphere, and this heat exchange plays a major role in the formation of atmospheric conditions, including atmospheric circulation and its cyclonic or anticyclonic nature. Salinity is also important as an indicator of the freshwater budget of adjacent areas. Hence, there is considerable interest in investigating the salinity of the Arctic Ocean and spatial-temporal patterns in its variability, especially in the surface layer where this variability is most strongly manifested.

Alekseev *et al.* (2000) calculated freshwater content relative to a referenced salinity of 34.8‰ (Aagaard and Carmack, 1989) and investigated the distribution of salinity in the upper 400-m layer of the Arctic Basin. The multiyear average freshwater content is maximum at the center of the anticyclonic Beaufort Gyre (approximately at $76°20'N$, $152°W$), where the freshwater content reaches 19 m, gradually decreasing to 1–2 m with increasing distance from the center of the gyre to its periphery. The origin of the Beaufort Gyre freshwater reservoir and its possible influence on Arctic Ocean circulation and climate are described by Proshutinsky *et al.* (2002), who show that the major cause of the large freshwater content in the Beaufort Gyre results from the process of Ekman pumping associated with climatological anticyclonic atmospheric circulation over the Canada Basin, centered in the Beaufort Gyre region.

Data available from many years of oceanographic observations, including broad surveys and data from drifting and polar stations, allowed us to calculate the mean annual freshwater content in the surface layer (up to 50 m), and obtain values of its multiyear changes within most of the Arctic Ocean. Because the amount of data available per unit area varied greatly in time and space, a technique for reconstruction of oceanographic characteristics was developed based on the spectral expansion method (Koltyshev and Timokhov, 1997). Using the oceanographic database of the Arctic Ocean, the temperature and salinity fields were reconstructed for March–May and a continuous series of characteristics was obtained for 1950 to 1993 at grid points with a spatial step of 200 km at standard oceanographic levels. The gridded fields of the reconstructed salinity values were used to calculate average salinity in the 5–50 m layer at the grid points, which made it possible to track changes in salinity for the indicated series of years (Ivanov *et al.*, 2003; Gudkovich *et al.*, 2004). According to these data, the freshwater content in the surface layer of the basin has decreased by approximately one-third over the 43-year time period.

Surface water salinity is known to be influenced by many factors:

— Freshening due to river runoff.
— The balance of atmospheric precipitation and evaporation at the ocean surface.
— The balance of the processes of ice growth and melting.
— Processes of upwelling and downwelling near the shores and landfast ice under the influence of winds.
— Processes of convection and vertical turbulent mixing of waters.
— Salt advection by currents of different origins.
— The influence on the Ekman pumping layer of non-uniform wind fields (baric fields—cyclones and anticyclones) accompanied by upwelling (in cyclonic systems) and downwelling (in anticyclonic systems).

Changes in the intensity and direction of these processes in time and space results in corresponding salinity changes. The large-scale processes appear to be of greatest interest for studies of climate change because they have major long-term consequences. The last three processes listed are the most important, especially the last, while the first four either have a lesser broad-scale influence or are local in character.

Figure 4.19a, b shows changes in the distribution of average salinity in the 5–50 m water layer in the Arctic Basin, described by linear trends for the periods 1950–1988 and 1989–1993 (Gudkovich *et al.*, 2004). At first glance, the distribution of salinity changes in space in the Arctic Ocean has a mixed character. However, some important repeated features are evident in these changes. During the time interval 1950 to 1988:

— A salinity increase was noted over much of the Arctic Basin, especially in the area of the Beaufort Gyre.
— The changes had the opposite sign (freshening) at the periphery of the gyre, in a zone extending from the north coast of Greenland to the New Siberian Islands.
— A salinity increase is also observed in the northern Greenland Sea and in the area adjoining the Barents Sea to the north, in the Pechora Sea, and in the northern Chukchi Sea.
— Freshening is also observed in the western Greenland Sea and in the Norwegian, Barents, Kara, Laptev, and East Siberian Seas.

A salinity decrease in the Arctic Seas (except for the Chukchi Sea) can at least be partly connected with increased runoff from large Asian rivers flowing to the Arctic Seas (the increase comprised approximately 10%). A salinity increase in the Chukchi and Beaufort Seas may result from corresponding changes in advection of Pacific Ocean waters. An increase in salinity over much of the Arctic Basin, especially in the Beaufort Gyre, is attributed to changes in atmospheric circulation: attenuation of the Arctic High and increased recurrence of cyclonic fields (see Section 4.2).

We remind readers that the upwelling that results in increased surface-water salinity in the central areas of baric depressions is accompanied by downwelling and freshening at their periphery. This probably explains the presence of a freshening belt extending from Greenland to the New Siberian Islands and the Siberian shelf seas (Figure 4.19a). Salinity change in the Greenland Sea is similar: deepening of the Iceland trough leads to salinification of surface water in the area known as the cold water dome, located in the northern part of the sea, and to freshening at the periphery of the cyclonic gyre in the Norwegian Sea (Gudkovich and Kovalev, 2002a). Intense freshening during the second half of the twentieth century was confirmed by Belkin *et al.* (1998) using direct-observation data taken onboard the weather ship (66°N, 2°E) in the southeastern Norwegian Sea.

As Figure 4.19b shows, during the next relatively short time interval (1989–1993), there were significant changes in the distribution of salinity trends in the surface layer of the Arctic Basin. A zone of increasing salinity moved westward, overlapping the area where freshening was previously observed. Salinity began to decrease in the area adjoining the north shores of the Canadian Arctic archipelago. A zone of freshening also appeared to the north of the Barents Sea. The main cause of these changes was significant modification of the pressure field expressed in a substantial decrease in atmospheric pressure in the region between Greenland and the Laptev Sea (Figure 4.20). As might be expected, salinification of surface waters in this region due to upwelling was accompanied by freshening at its periphery.

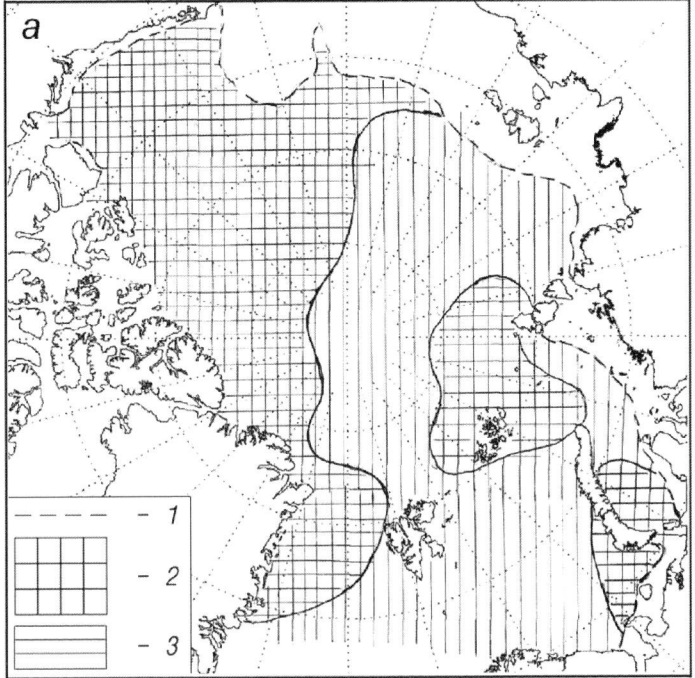

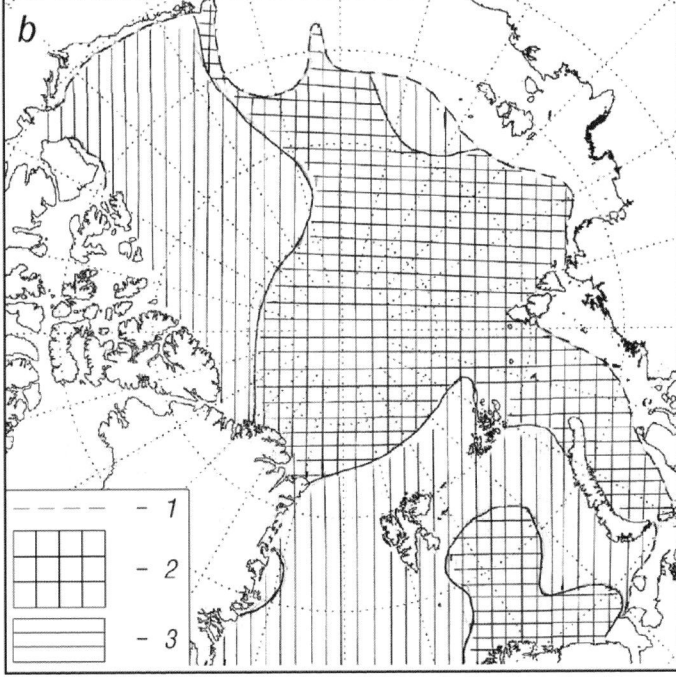

Figure 4.19. Changes in the distribution of average salinity in the 5–50-m water layer described by the linear trends for the periods 1950–1988 (a) and 1989–1993 (b). (1) 100-m isobath. (2) Salinity increase. (3) Salinity decrease.

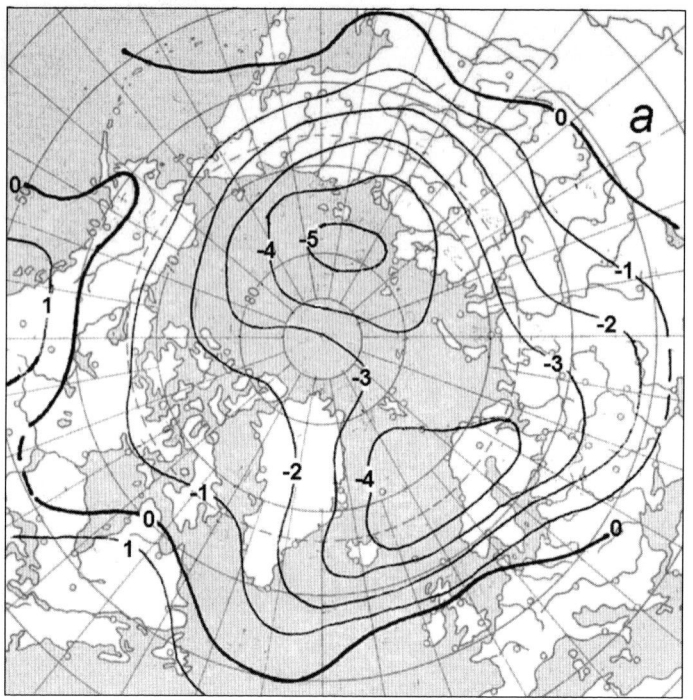

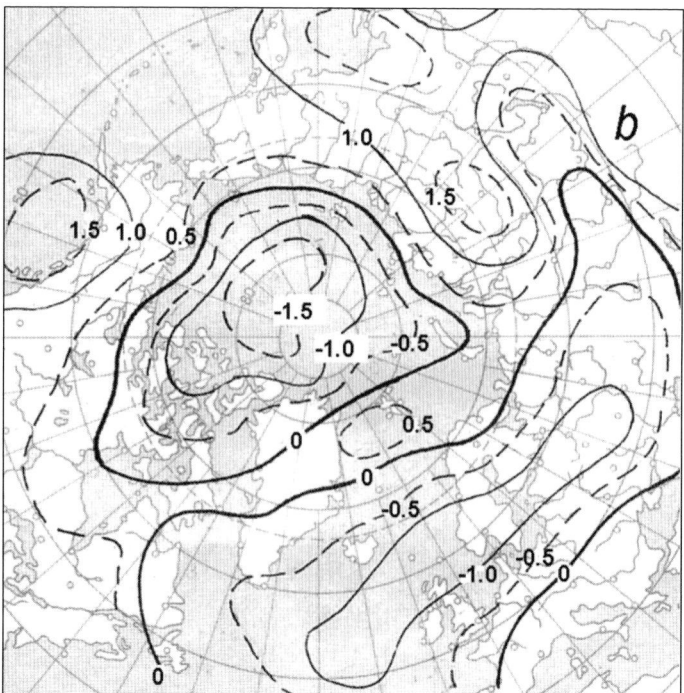

Figure 4.20. Average difference in atmospheric pressure (hPa) between the periods 1985–1995 and 1970–1980 for January–March (a) and July–September (b).

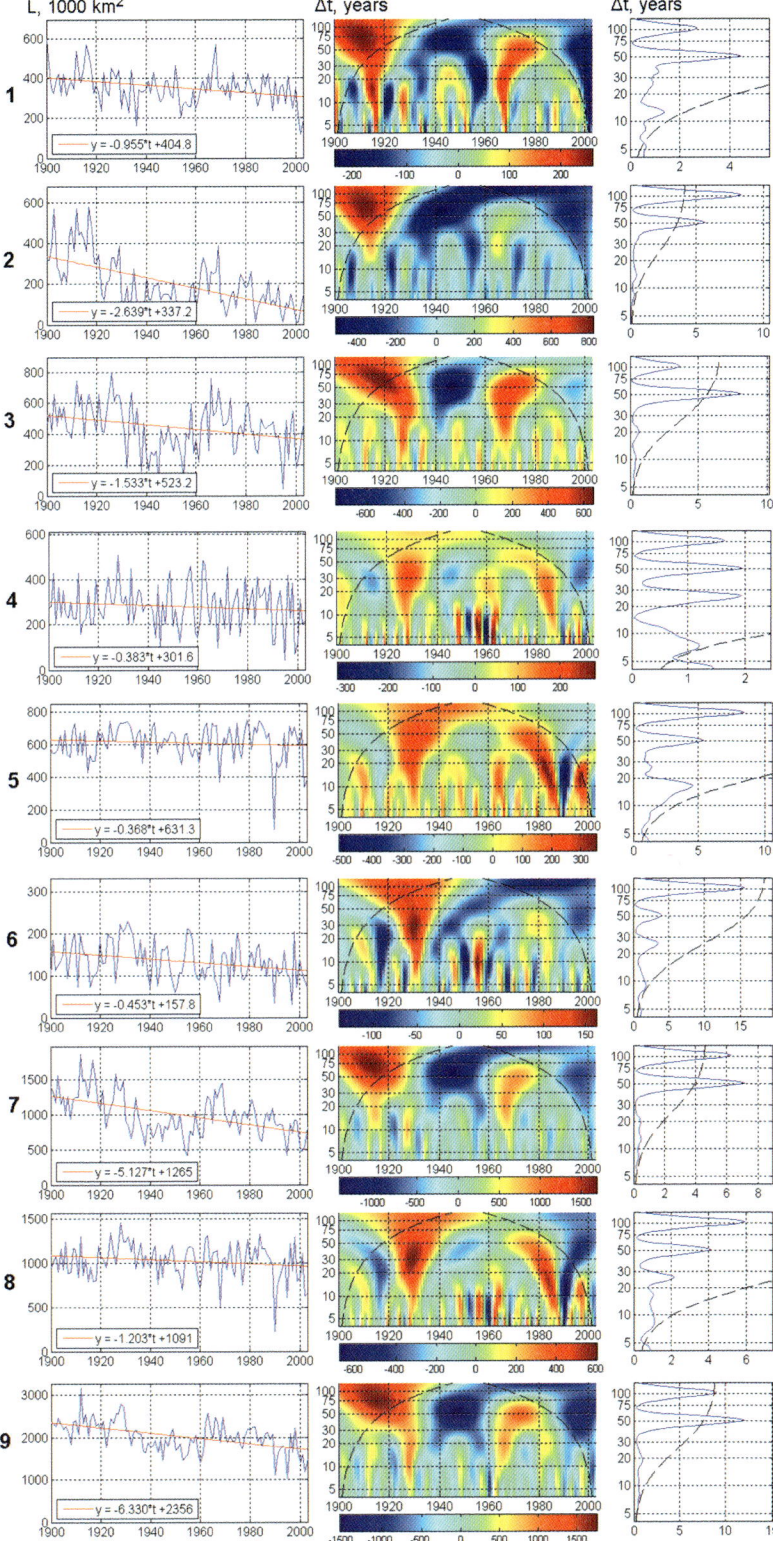

Figure 2.6. Temporal variability of the spectral structure of ice extent fluctuations of the Arctic Seas in August for 1900–2003 on the basis of wavelet-analysis data: 1) the Greenland Sea, 2) the Barents Sea, 3) the Kara Sea, 4) the Laptev Sea, 5) the East-Siberian Sea, 6) the Chukchi Sea, 7) the western sector seas (1–3), 8) the eastern sector seas (4–6), 9) the Eurasian Arctic seas (1–6); a) time series, b) wavelet spectrum, c) total wavelet spectrum. The insets in (a) show equations for linear trends and the dashed lines in (b) and (c) denote the area of 95% statistical probability.

(a) **H, 1000 km²**

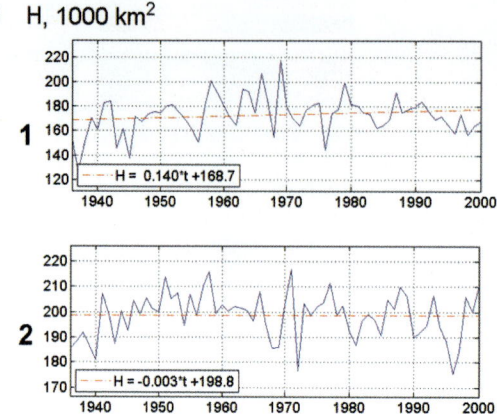

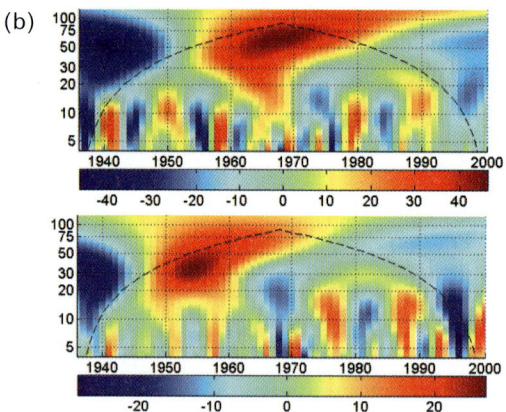

(b)

(c) **ΔT, years**

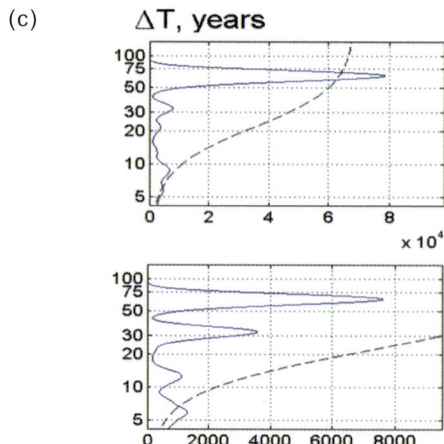

Figure 3.1. Temporal variability of maximum landfast ice thickness for 1936–2000 and its spectral structure based on wavelet data analysis. (1) Kara Sea stations. (2) Laptev, East Siberian, and Chukchi Sea stations. (a) Time series. (b) Wavelet spectrum. (c) Total wavelet spectrum. The insets in (a) show the equations for linear trends. The dashed lines in (b) and (c) show areas of 95% statistical probability.

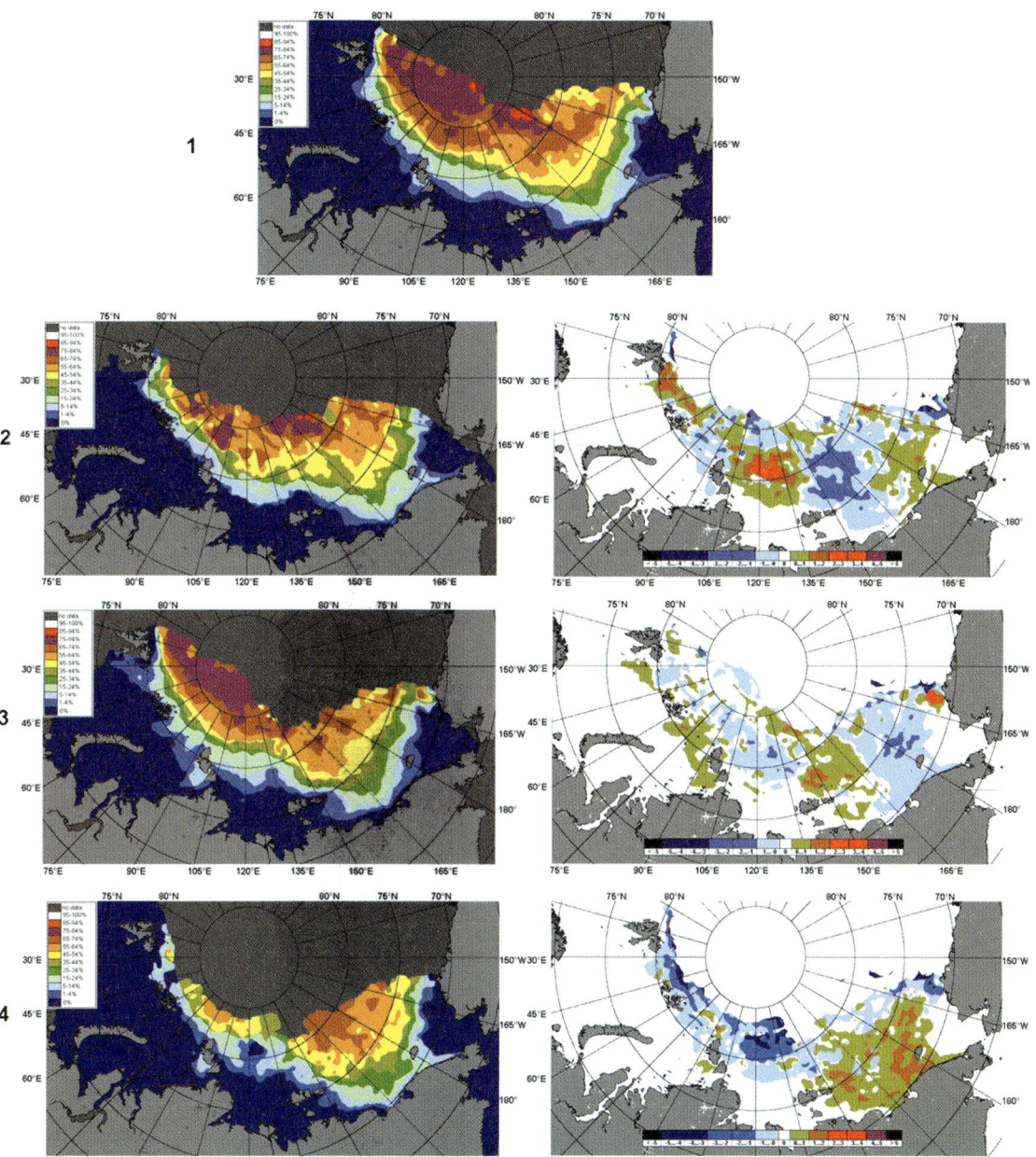

Figure 3.2. Left column: Statistical parameters of old-ice partial concentration in August for 1933–1992 and 1940–1959, 1960–1979 and 1980–1992. Right column: Differences: (2) 1940–1959 compared to 1933–1992, (3) 1960–1979 compared to 1933–1992, (4) 1980–1992 compared to 1933–1992.

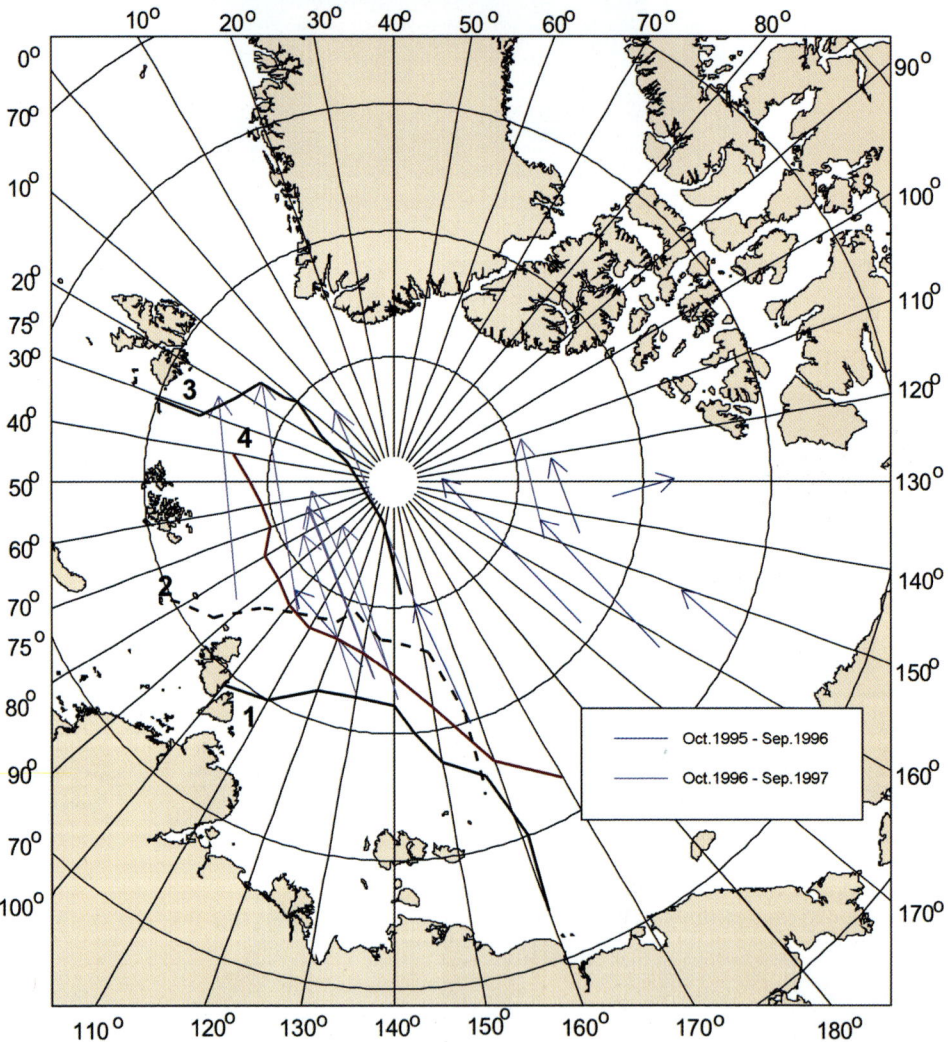

Figure 3.3. (1) Location of the residual ice edge at the end of September 1995. (2) Calculated location of the second-year ice extent boundaries at the end of September 1996, and (3) multiyear ice at the end of September 1997. (4) Mean climatologic boundary of multiyear ice in September. Blue arrows indicate the drift of IABP buoys from October 1995 to September 1996. Red arrows indicate the drift of IABP buoys from October 1996 to September 1997.

01.06–15.06

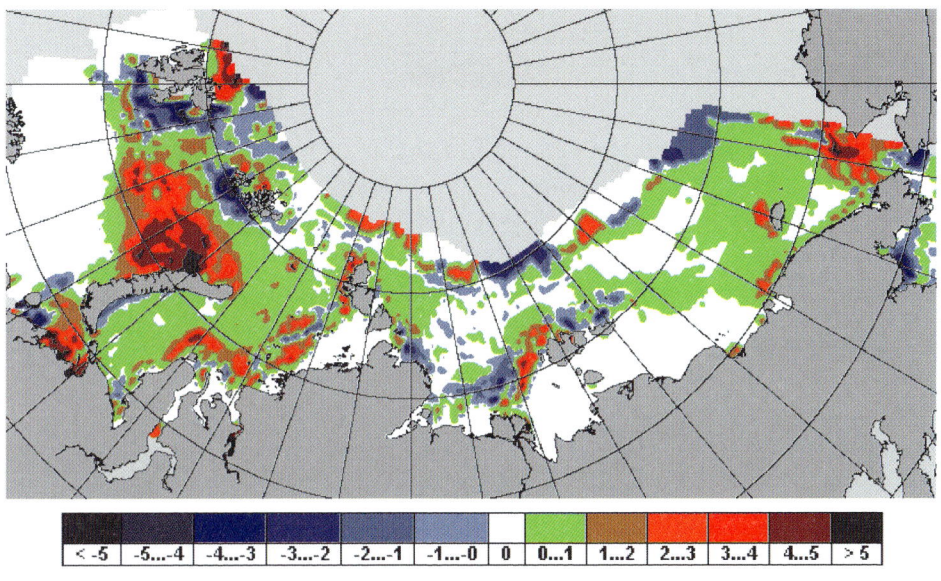

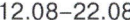

12.08–22.08

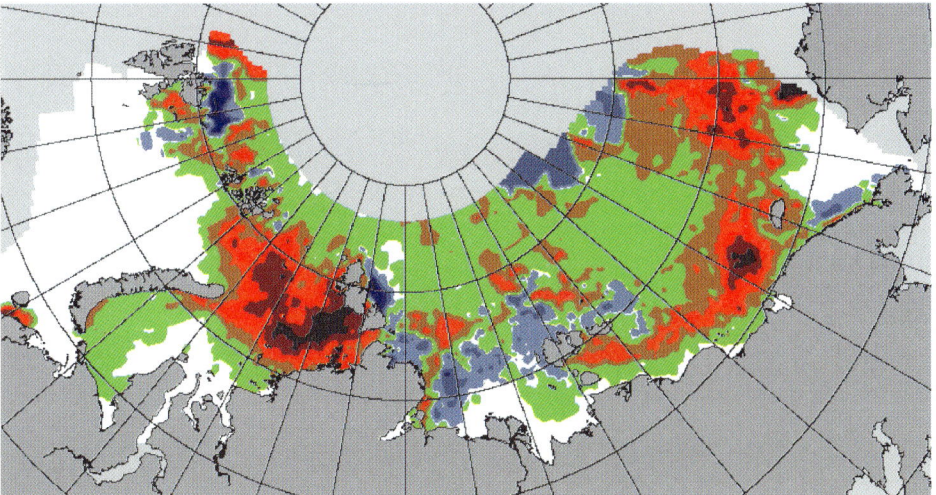

Figure 3.5. Distribution in space of the difference in total sea ice concentration averaged over the periods 1963–1983 and 1940–1962.

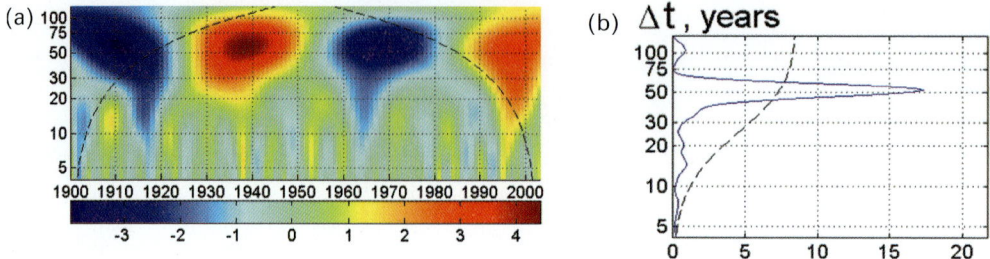

Figure 4.2. Temporal variability of the spectral structure of mean annual air temperature anomalies in the 70–85°N zone in the twentieth century based on wavelet-analysis results. (a) Wavelet spectrum. (b) Total wavelet spectrum. Dashed lines denote the area of 95% statistical probability.

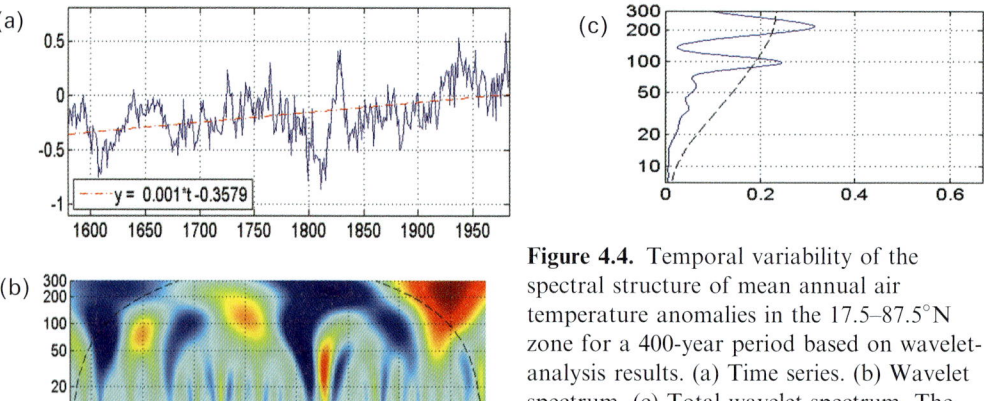

Figure 4.4. Temporal variability of the spectral structure of mean annual air temperature anomalies in the 17.5–87.5°N zone for a 400-year period based on wavelet-analysis results. (a) Time series. (b) Wavelet spectrum. (c) Total wavelet spectrum. The inset in (a) shows the equation for a linear trend. The dashed lines in (b) and (c) denote the area of 95% statistical probability.

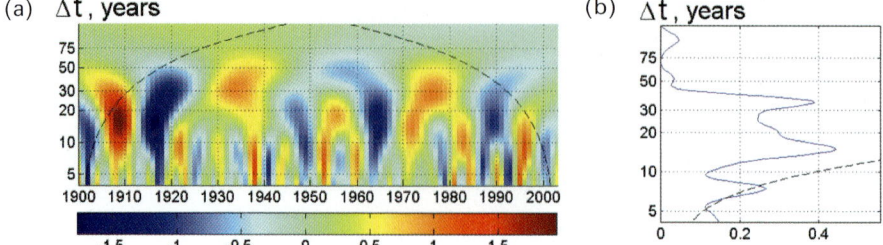

Figure 4.6. Temporal variability of the spectral structure of annual air temperature anomalies in the 70–85°N zone (note that linear trend and 50–60 year fluctuations are excluded) based of wavelet-analysis results. (a) Wavelet spectrum. (b) Total wavelet spectrum. Dashed lines denote the area of 95% statistical probability.

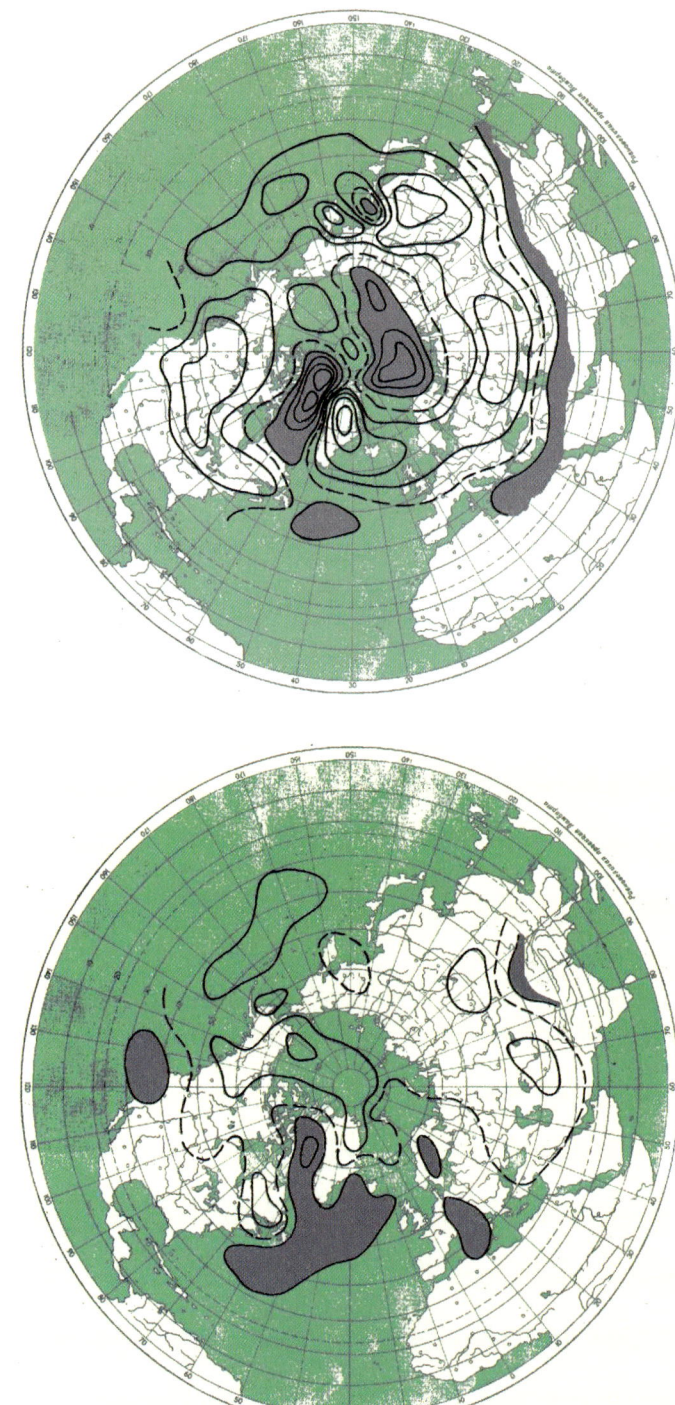

Figure 6.4. Differences in mean surface air temperature for 1980–2000 compared to 1930–1950 (a) from October to March, and (b) from April to September. Isolines are drawn at 1° intervals. Gray shading indicates areas with a difference of less than −1°.

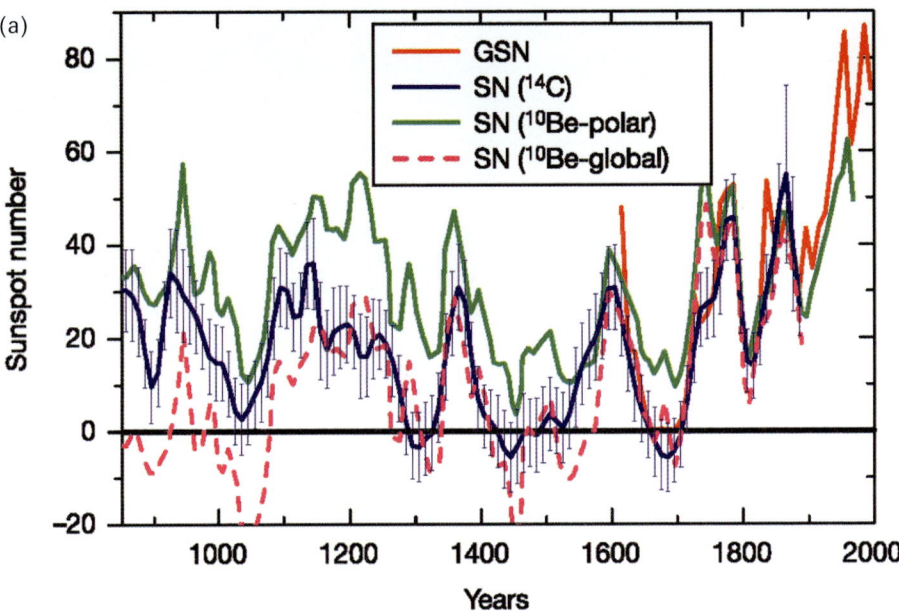

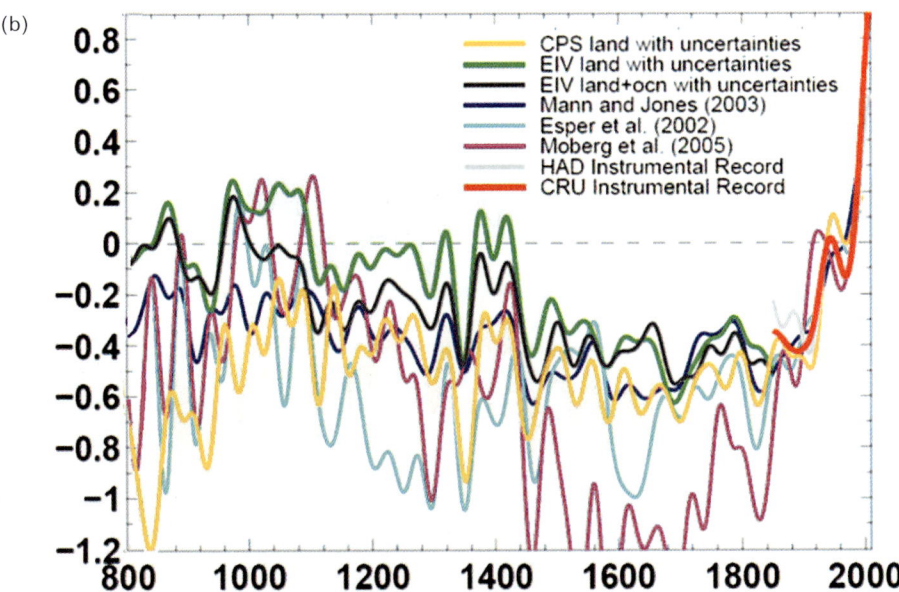

Figure 6.6. (a) Sunspot numbers averaged over 10 years from direct measurements (red line) and reconstructed from cosmogenic isotopes ^{14}C (blue line) and ^{10}Be (green and dashed magenta lines) (Solanki *et al.*, 2004). (b) Various reconstructions of Northern Hemisphere surface temperature (Mann *et al.*, 2008). Note that the sharp upward red line at the far right exaggerates 20th century warming beyond reality.

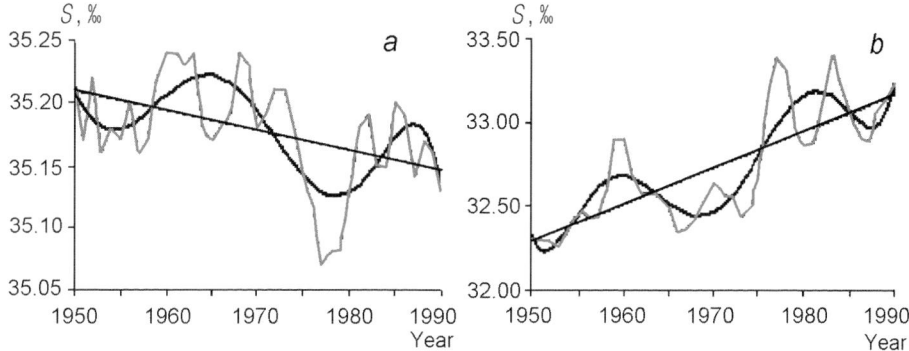

Figure 4.21. Change in average salinity in the 0–100-m layer in the Norwegian Sea from weather ship M observations (a) and in the 0–50-m layer in the Arctic Basin (b).

As described above, the multiyear series of salinity values in the surface water layer (5–50 m) of the Arctic Basin in the March–May period at regular grid points allowed us to calculate average values (S_{avg}) for the entire basin. Figure 4.21b plots changes in the S_{avg} value from 1950 to 1990, with the actual fluctuations approximated by a polynomial to the power of 6; there is a pronounced linear trend pointing to a gradual salinity increase in the basin from the beginning to the end of the time interval. Cyclic changes are also identified:

— Increase in salinity in the late 1950s–early 1960s.
— Subsequent decrease in salinity in the late 1960s–early 1970s.
— A new increase in salinity around 1980, which was replaced by its decrease at the end of the time series.

The period of observed cyclicity lasts about 20 years. There are also higher frequency fluctuations with periods of 6–10 years.

It is reasonable to compare the salinity changes in the surface water layer of the Arctic Basin with the corresponding processes in the Norwegian and Greenland Seas, where salinity fluctuations of the same frequency were revealed and explained by the presence of self-oscillations in the ocean–ice cover–atmosphere system (Gudkovich and Kovalev, 2002a) (also see Section 5.3).

Figure 4.21a plots salinity changes within the 100 m horizon for the Norwegian Sea from weather ship M observations. Comparison of Figures 4.21a and b indicates a surprising interrelationship between the two areas. Because observations onboard the weather ship in the Norwegian Sea were made at the periphery of the cyclonic gyre, where the water density changes are opposite in sign to the changes occurring in the central zone of the gyre (Gudkovich and Kovalev, 2002), it can be concluded that long-term salinity changes in both regions had a similar character. Salinity fluctuations in the surface layer of the Arctic Basin exhibit a 3.25-year lag, on average, compared to the changes at the center of the cyclonic gyre of the Greenland Sea (Table 4.7).

Table 4.7. Years of salinity maximums and minimums in the upper water layer, vorticity of the wind fields, and corresponding lag values in two regions of the Arctic Ocean, the Greenland Sea (GS), and the Arctic Basin (B).

Extremes	Salinity (S)			Vorticity of the wind fields (ΔP)			τ, years $(\Delta P - S)_{AB}$
	GS	AB	τ, years $(AB - GS)$	GS	AB	τ, years $(AB - GS)$	
Maximums	1954	1959	+5	1952	1948	−4	11
Minimums	1964	1968	+4	1962	1958	−4	10
Maximums	1978	1982	+4	1976	1969	−7	13
Minimums	1987	1987	0	1985	1980	−5	7
Maximums	1997	–	–	1995	1990	−5	–
Average			+3.25	–	–	−5.0	+10.25

It is useful to estimate the time lag of average salinity variations in the surface layer of the Arctic Basin relative to the baric field changes over it. As an indicator of the latter, the large-scale vorticity index values were calculated using the Laplacian of mean atmospheric pressure at $\varphi = 84°$N, $\lambda = 130.2°$E:

$$J_0 = \sum P_i - 4P_0 \tag{4.11}$$

where P_i, P_0 are atmospheric pressure at the points located at the square angles and at its center, respectively for Region 0 (see Figure 4.22).

Figure 4.23 shows a change in time for the average values of this index for March–July smoothed by running 11-year periods. There is a clear 20-year cycle in the salinity variations in this index. Fluctuations in this index are generally close in phase to the changes in the high-latitude zonality index (I_z) considered above (Figure 4.7): the location of the extremes differs within ± 2 years.

North Atlantic cyclones are known to become significantly deeper in the region of the Norwegian Energy Active Zone (NEAZO), which is connected to an important center of atmospheric circulation, the Icelandic depression. Hence, cyclones spread to the northeast and east, influencing the formation of the baric field over the Arctic, northern Europe, and Asia.

When the cyclones move to the northeast (Vangengeim–Girs circulation—type E), a baric trough formed over the Arctic Basin and the seas from the Barents to the Laptev becomes deeper with increased cyclonic activity in NEAZO, and is partly filled upon weakening. By slightly simplifying the processes, it can be concluded that vorticity of the wind fields in the area of the Icelandic depression and in the Arctic should change quasi-synchronously or with a small lag along cyclone pathways.

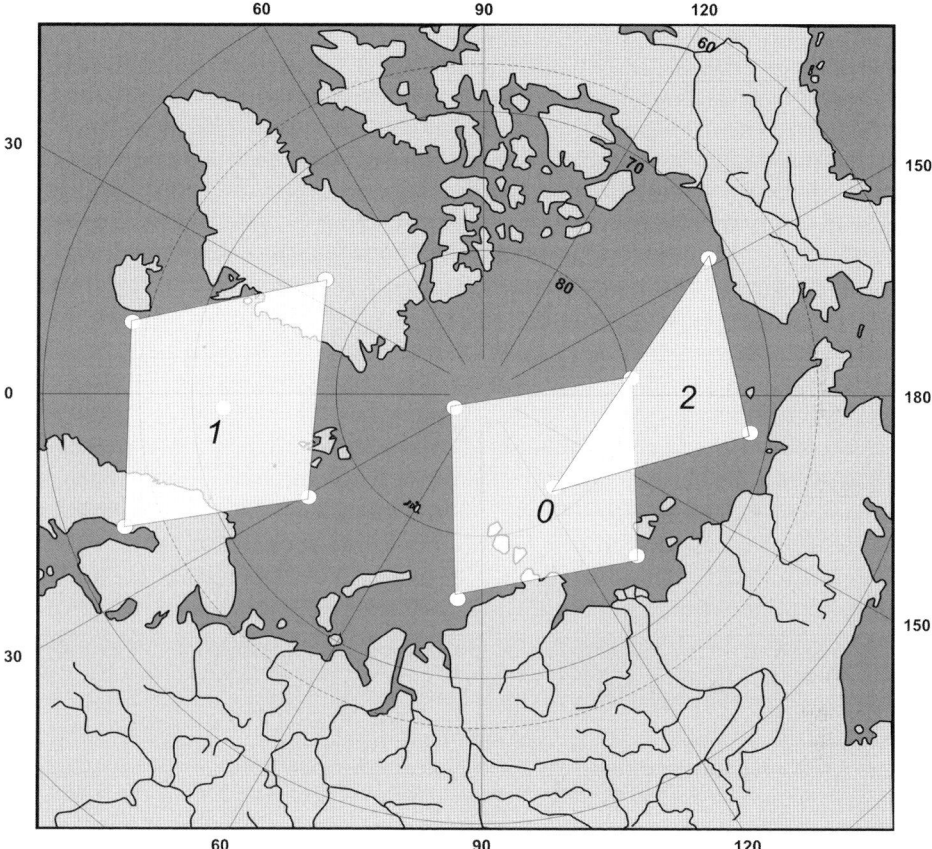

Figure 4.22. Layout of the regions for calculation of vorticity index J_0, J_1, and J_2 values.

When cyclones move to the east (circulation type W), the baric trough is mainly formed over northern Europe and northern Asia, and the Arctic High should intensify. Hence, changes in wind-field vorticity and corresponding atmospheric pressure anomalies in the Arctic and in the region of the Icelandic depression will mainly occur in opposite phase. Under real extended conditions, there is usually cyclone movement both to the east and northeast (there are only changes in recurrence of different trajectories). Thus, some phenomena occur during changes in wind-field vorticity and corresponding atmospheric pressure anomalies, when processes in the Arctic Basin precede changes in the Iceland depression area. Exactly such a situation is apparent in a comparison of maxima and minima in 20-year cyclic changes in baric fields (Table 4.7).

Table 4.7 shows that the maxima and minima of cyclonic activity in the Arctic Basin precede (on average by 5 years) similar maxima and minima in the Greenland Sea. If the salinity response in the Greenland Sea to atmospheric conditions has an average lag of 2 years, the same response in the Arctic Basin takes about 10 years. The

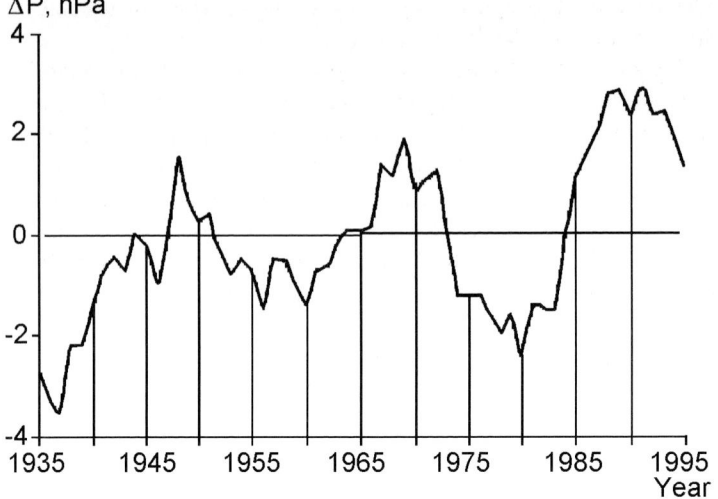

Figure 4.23. Variation over time of the vorticity index value of $J_0 = \Delta P$ at the point of $\varphi = 84°N$, $\lambda = 130.2°E$ for March July using 11-year running averaging.

differences must be caused by differences in the intensity of atmospheric processes in these regions. This result confirms the earlier conclusion by Gudkovich and Nikiforov (1965) regarding the significant stability of large-scale water circulation in the Arctic Basin.

The results of the present study show that 20-year cycles of change in the Greenland Sea generate corresponding changes in the Arctic and probably in the North Atlantic where a 21-year cycle is evident in the spectral density function of the NAO index (Figure 4.11). The trends in multiyear large-scale processes in the atmosphere and the ocean provide evidence of the influence of natural and possibly of anthropogenic factors or of cyclic fluctuations lasting longer than 100 years. Both factors may be at work here, along with processes taking place in the energy-active zone of the Greenland Sea.

Swift *et al.* (2005) studied the variability of different Arctic Basin water masses during the second half of the twentieth century by subdividing the basin into 13 boxes and averaging the water properties in each box. Based on their findings, these authors posed alternative hypotheses to the explanation of the main cause of surface water salinification.

As noted above, surface water salinity in some seas depends on their sea ice extent. The inverse character of the relationship between sea ice extent and salinity was confirmed by observations in the Kara and Chukchi Seas (Gudkovich *et al.*, 1972, 1997). This relationship was observed in spite of the fact that the water–ice phase transitions should have resulted in the opposite changes: increased growth and decreased ice melting in the "cold" epochs should have resulted in salinification, and the decreased growth and increased melting in the "warm" epochs should have had the opposite effect. As shown in Appel and Gudkovich (1984), salt advection by ocean currents has a much greater influence on salinity (for example, the flow of relatively saline Barents Sea water to the Kara Sea through Makarov Strait and transport of Pacific Ocean water through Bering Strait to the southwestern Chukchi Sea with the Long Strait branch of the Bering Sea current).

These patterns disprove the widespread opinion expressed in scientific publications that the Arctic warming that began at the end of the twentieth century is accompanied by freshening of Arctic Ocean surface water, increased outflow of Arctic Ocean surface water to the North Atlantic, and a corresponding influence of Arctic Ocean surface water on thermohaline circulation in this region (e.g. Hassol, 2004, etc.). As demonstrated above, due to a weakened Arctic High (increased cyclonic activity in the Arctic) and the inflow of more saline oceanic waters from the south, the salinity of surface water over much of the Arctic Basin during this period has increased. Ice export from this basin to the Greenland Sea has decreased (see Section 4.4.1). This could lead not to freshening but rather to salinification of North Atlantic water. We think that the area of decreasing salinity in the North Atlantic has been limited to the northwestern region of the Atlantic Ocean adjoining Davis Strait, located behind the Icelandic depression.

The peak for river runoff and ice formation and melting processes in the Norwegian Sea (minimum salinity percentage) is known to have occurred at the end of the 1970s (see Figure 4.21a). At the end of the period of Arctic cooling, ice export from the Arctic Basin to the Greenland Sea intensified, resulting in freshening of the water in the northeast Atlantic and in the North European Basin. Thus, descriptions of the processes related to climate warming at the end of the 20th century were significantly misinterpreted by those who claim that this natural phenomenon will prove to be catastrophic if left unchecked.

Water temperature is also quite significant in the processes of climate change. Temperature profiles of different Arctic Basin water masses are presented in studies by Timofeyev (1960), Frolov *et al.* (2005), Polyakov *et al.* (2004, 2005), and others. In the surface layer of the Arctic Basin, the water temperature is close to the freezing point of water of relevant salinity. However, in the deep layers that contain relatively warm water of Atlantic origin, the temperature depends on both the volume and temperature of incoming Atlantic water. The effects of climatic change on these parameters were observed by Zubov (1938), who noted that during the period of Arctic warming in the 1920s–1930s, the average water temperature of the Nordkapp current (0–200 m layer) was 0.7–0.8°C higher than at the beginning of the twentieth century. Based on observations of the 1937–1940 expedition aboard the icebreaker *G. Sedov*, Timofeyev (1960) showed that the average temperature in the Atlantic water layer of the Arctic Basin was much higher than that measured in the same region by the 1893–1896 *Fram* expedition, while the maximum temperature of this layer increased by 0.7–1.0°C. There were significant interannual fluctuations in the mid-twentieth century as the temperature began to decrease. This was confirmed by Bulatov and Zakharov (1978), who investigated changes in the thermal state of the Arctic Ocean for a 20-year period from the mid 1950s to the mid 1970s. These authors observed some cooling of Arctic Basin waters, consistent with atmospheric cooling during this period. This cooling was more pronounced in the sub-Atlantic sector of the basin than in its sub-Pacific Ocean sector. Lamb and Johnson (1964) found that the temperature of deep Atlantic water was even lower at the peak of the Little Ice

Age (1790–1829), when the water temperature at the surface of the North Atlantic was 2–3°C lower than the current temperature.

Alekseev *et al.* (1998) focused their study on a significant new increase (anomaly of about +1°C) in the deep Atlantic water temperature in the 1990s that had been previously revealed by Quadfasel (1991) and Schauer *et al.* (1995). The upper boundary of this water mass was significantly higher, and the layer was thicker. An analysis of these observations suggests that each twentieth-century cycle of climate warming was accompanied by an increase in the temperature and the heat content of the deep Atlantic layer in the Arctic Basin. The quantitative indicators of warming of this water provide evidence that the 1990s warming event was similar to the 1930s Arctic warming. Note that comparing the anomalies of Atlantic water temperature in different years requires accounting for the location of the main flow of this water near the Eurasian continental slope, where the anomalies of climatic changes are maximized; they decrease rapidly toward the Canadian Arctic archipelago (Alekseev *et al.*, 1998).

Alekseev *et al.* (1998) identified a significant phenomenon that accompanies warming: an increase in salinity of the upper water layer. This results in a decrease in the vertical density gradient and a corresponding increase in the heat flux from depth, which can contribute to a decrease in sea ice thickness along with an increase in air temperature and snow cover thickness during epochs of climate warming. As noted in Section 3, the intensity of ice growth observed onboard *G. Sedov* was 20% less than that observed during the *Fram* drift.

In an interesting study, Polyakov *et al.* (2004) analyzed a large set of observational data on the changes in Atlantic water temperature in the Arctic Basin during the twentieth century and found that low-frequency fluctuations with a period of about 60 years are clearly evident (Figure 4.24). Coherence is shown in the fluctuations of Atlantic water temperature and surface air temperature, ice cover in the Arctic Seas, and thickness of landfast ice in the vicinity of polar stations. The increased inflow of warmer Atlantic water through the Norwegian Sea is accompanied by changes in water density distribution in the Arctic Basin, which, along with the increased outflow of cold and freshened water to the North Atlantic through Davis Strait, reduces low-frequency fluctuations in the atmosphere–ocean–ice cover system.

Most researchers who have investigated changes in temperature and other properties of deep Atlantic water in the Arctic Basin correlate these changes with modification of atmospheric circulation (increase in its intensity and recurrence of cyclone penetration to high latitudes). An indirect confirmation of the increased inflow of Atlantic water to the Norwegian Sea and farther north is provided by the aforementioned decrease in surface water layer salinity in the eastern part of the sea as a result of adaptation of the field of masses to the system of currents. However, surface water freshening is accompanied by the increased stability of water masses, which leads to a decreased rate of deep-water cooling. As a result, the temperature of the water flowing to the Arctic Basin increases. This is confirmed by the results of isotopic analysis of the seabed (Duplessy, 1980; Flohn, 1980), showing that during glacial epochs, when vertical circulation was restricted by a

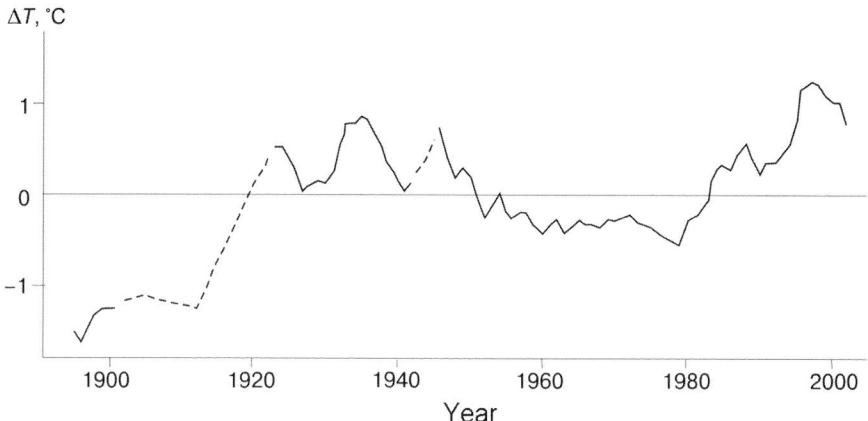

Figure 4.24. Long-term variability of Atlantic water temperature in the Arctic Basin in the twentieth century (Polyakov *et al.*, 2005). Dashed line is extrapolation for missing data.

comparatively thin surface layer, the near-bottom water temperature in the Norwegian Sea was higher than during the current epoch.

To determine the influence of the distribution of water masses (dynamic heights) in the Arctic Basin on sea ice extent of the Asian-shelf Arctic Seas, Koltyshev and Timokhov (1997) used expansions of the fields of dynamic heights and sea ice extent by EOF and by singular values (SV). The cross-correlation analysis allowed them to determine that the sea ice extent of some marginal seas depends on the structure of dynamic height fields during the preceding significant time intervals (from several months to two years). An inverse impact of sea ice extent on water circulation in some seas was also revealed and attributed to dependence of surface water salinity on sea ice extent. Koltyshev and Timokhov (1997) speculate that sea ice and ocean changes are organized as a self-oscillating system with a period of approximately 6 years.

4.7   LONG-TERM CHANGES IN RIVER RUNOFF

To some extent, long-term changes in river runoff, which contribute to the formation of a system of surface currents in the Arctic Ocean and in the upper low-salinity water layer in the Arctic Basin, can influence fluctuations in sea ice area and distribution.

Zakharov (1996) compares the volume of continental runoff to the Arctic Ocean from Asia and North America for 1940–1968 (Anon. (N)) with the extent of sea ice in the North European Basin from 1946 to 1971 (mean annual data). The correlation coefficients of this relationship at time shifts of 3, 4, and 5 years (the sea ice extent after the runoff) were 0.33, 0.45, and 0.36, respectively (at a 95% significance level of 0.51). However, with 5-year running smoothing of the series and a shift of 2 years, the coefficient value increased to 0.82. We suggest that this relationship cannot provide a convincing argument in favor of the decisive role of continental runoff in sea ice extent changes because, as Zakharov (1996) indicates, the data were obtained using indirect methods. The anomalies of iceberg discharge were not taken into account, and the series compared are short (25 years). The continental runoff that was taken into account comprises only 42% of the freshwater flowing into the Arctic Ocean. A similar correlation of sea ice extent (for a longer period, 1940 to 1999) using observational data on runoff of the largest rivers to the seas of the Arctic Basin, which feed freshened Arctic water directly to the North European Basin, does not confirm this sea ice extent relationship to continental runoff. This author proposes a conceptual scheme of self-oscillation in the atmosphere-ocean-polar ice system that also provokes some strong objections (see Section 5.3).

Based on data in Ivanov (1976) and Zakharov (1996), the total continental runoff to the Arctic Ocean is 5135 km^3/year. Ivanov *et al.* (2004) estimate continental runoff to the Russian Arctic Seas at approximately 2900 km^3/year, including 2300 km^3/year delivered by the large rivers flowing to the Eurasian seas. River runoff is non-uniformly distributed during the year: in summer (May to October), it comprises 84–85% (in the Barents and Kara Seas) to 99.5% (in the Chukchi Sea). But even in such rivers as the Yenisey with a significant part of their drainage area located outside the permafrost zone, almost half (45%) of the annual runoff is observed during the flooding that occurs in one month (June). Hence, runoff to the Arctic Seas mainly depends on the accumulation of solid precipitation in the winter. The intensity of this process depends on the speed of zonal transport in the troposphere of temperate latitudes that brings relatively warm and moist air from the North Atlantic.

The temporal changes in river runoff to the Eurasian Arctic Seas are characterized by the presence of relatively short-term cyclic fluctuations with durations of 3–5 years (the White, Barents, and Chukchi Seas), 5–6 years (the Laptev Sea), 6–8 years (the East Siberian Sea), and 8–12 years (the Kara Sea) (Ivanov *et al.*, 2004). An analysis of multiyear changes in the annual runoff of large rivers to these seas for 1937–1999 (using data kindly provided by Ivanov *et al.* (2004)) showed the presence of a noticeable positive trend (except for the Kolyma, where a small negative trend is observed). The long-term changes in river runoff to the Arctic Seas throughout much of the twentieth century are shown in Figure 4.25.

As shown in Figure 4.25, the parameters of linear trends of annual runoff to the western seas (the Barents and Kara) and to the eastern seas (the Laptev and East Siberian) differ significantly. Of interest are typical changes in the river runoff trends in 1967 (western seas) and about 1973 (eastern seas). The trend parameters are shown in Table 4.8.

River runoff to the Eurasian shelf Arctic Seas increased significantly in the last third of the twentieth century (Figure 4.25 and Table 4.8). This phenomenon was caused by intensified west-to-east circulation in the atmosphere of middle and temperate latitudes in the Northern Hemisphere, which is associated with a corresponding decrease in atmospheric pressure at high Arctic Ocean latitudes. The North Atlantic Oscillation (NAO) index is discussed in Section 4.2 as a good indicator of the intensity of North Atlantic west winds; it also reflects a general planetary west-to-east transfer at temperate latitudes in the Northern Hemisphere (Smirnov *et al.*, 1998).

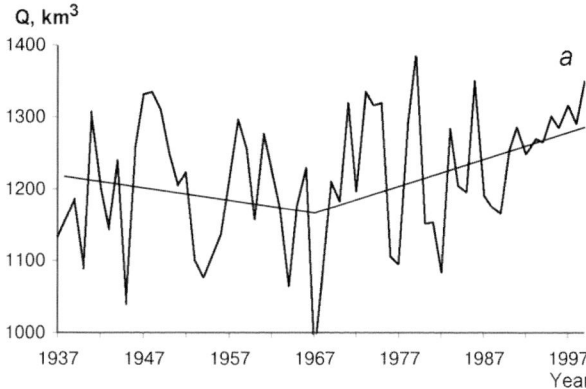

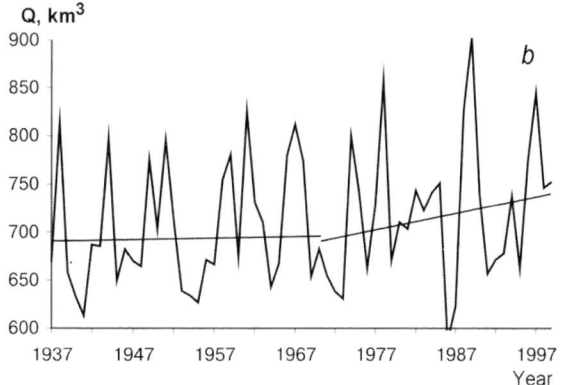

Figure 4.25. Changes in the total annual runoff of the Severnaya Dvina, Pechora, Ob', and Yenisey Rivers (a) and the Lena, Yana, Indigirka, and Kolyma Rivers (b) from 1937 to 1999. Straight-line segments indicate linear trends for the typical time intervals.

Table 4.8. Linear trends (km³/year) in river runoff by region and time

Western Seas Region		Eastern Seas Region	
Period, years	*Trend*	*Period, years*	*Trend*
1937–1967	−1.34	1937–1973	+0.24
1967–1999	+4.00	1973–1999	+1.10

Figure 4.26 shows fluctuations in the NAO index for the period 1937 to 1994. It depicts linear trend segments that approximately correspond to two time intervals considered above. The parameters were −0.057 and +0.122, which is in satisfactory agreement in sign with the changes in trend parameters of the river runoff volume to the western seas.

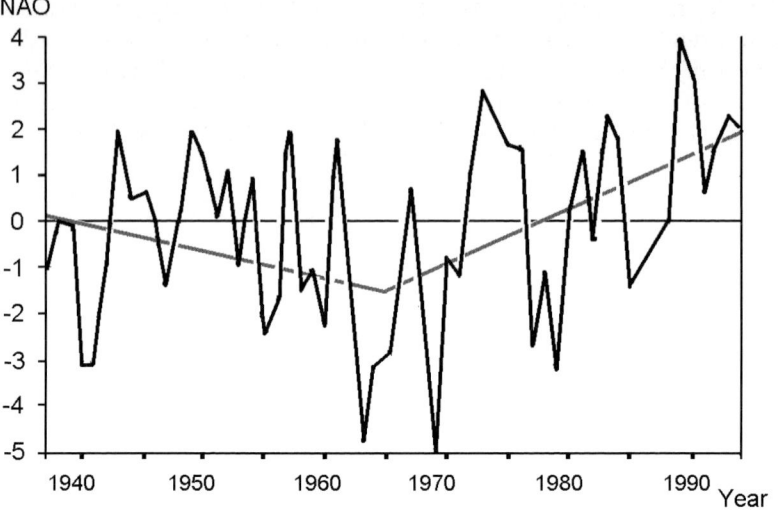

Figure 4.26. Fluctuations in the winter North Atlantic Oscillation (NAO) index for the period 1937–1994.

Increased west-to-east transports in the atmosphere of temperate latitudes at the end of the twentieth century are confirmed by our zonality index, which characterizes the intensity of these transports between 40° and 65°N, as shown in Figure 4.27, which is a display of part of Figure 4.9. Comparison of the two figures shows that the linear trends plotted in Figure 4.27 mainly represent branches of a cyclic fluctuation with a period of about 60 years. The elevated runoff and the increase in the indexes at the end of the twentieth century compared with their values in the 1930s–1940s are probably determined by cycles lasting more than 100 years.

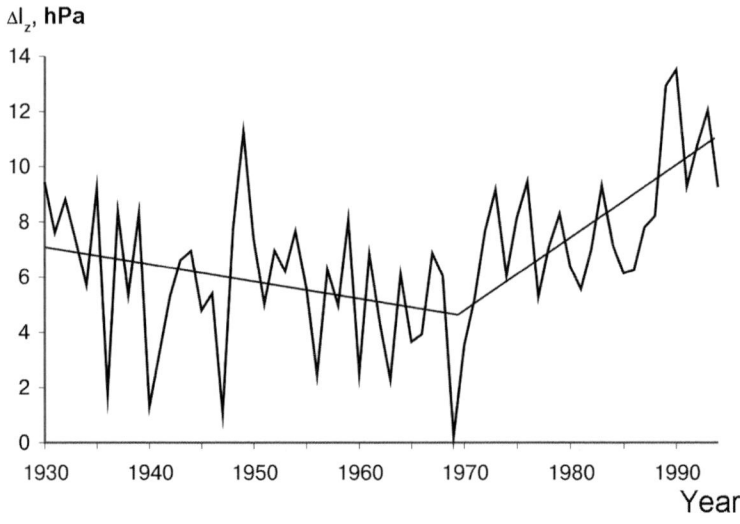

Figure 4.27. Changes in the average zonality index for October–March for the period 1930–1994.

Clear similarities in the trends of zonal transports in the atmosphere (Figures 4.26 and 4.27) and runoff of the Eurasian rivers to the Arctic Ocean (Figure 4.25) confirm the validity of the proposed hypothesis about the causes of climatic change, namely: [increased west-to-east transports in the moderate latitudes] → [increased precipitation over the drainage area] → [increased Siberian river runoff into the Arctic Basin].

5

Possible causes of changes in climate and in Arctic Seas ice extent

Understanding the causes of climate change at different time scales is still at the stage of framing scientific hypotheses, and hence requires further detailed investigation. Unfortunately, since climate change is by definition, a long-term phenomenon, it is very difficult to prove or disprove hypotheses. We have an abundance of hypotheses and a dearth of detailed long-term data. Nevertheless, where data exist, we should prefer data to computer models. Most cyclic and secular variations in sea ice conditions are rooted in atmospheric and oceanic processes that are influenced by both external and internal factors. The external factors include such helio- and geophysical impacts as solar activity, tidal and nutation phenomena, Earth's rotation speed, fluctuations in the solar constant due to changes in distance between the Earth and the Sun, fluxes of energy and charged particles from space, and other astronomical factors. Internal factors encompass natural hydrometeorological, geological, and biological processes and self-oscillation phenomena related to interactions in the lithosphere–ocean–sea ice–atmosphere–land system, with the latter including glaciers, rivers, etc. Anthropogenic factors or impacts that may augment internal system variables are associated with increased concentrations of greenhouse gases in the atmosphere and generation of black carbon soot and sulfate aerosols, due to human activities and their putative impact on climate.

5.1 ON THE QUESTION OF ANTHROPOGENIC IMPACT ON SEA ICE EXTENT VARIABILITY

Since the 1960s, an increasing number of climatologists have become concerned that the impact of greenhouse gases accumulating in the atmosphere as a result of the burning of hydrocarbon fuels and of deforestation may produce disastrous climate change in the 21st century. The report of the Intergovernmental Panel on Climate

Change for 2007 (IPCC, 2007) estimates the concentration of carbon dioxide (CO_2) in the atmosphere at 383 parts per million (ppm) in 2007, 37% over pre-industrial revolution concentrations (280 ppm in 1750), and higher than any concentration found in polar ice core records of atmospheric composition dating back 650,000 years—and possibly higher than any concentration in the last 20 million years. It is widely believed that the increased concentration of greenhouse gases reduces the Earth's long-wave emission, resulting in increased temperature in the troposphere and the surface. The basis for this belief rests mainly on two factors: (1) the global average temperature increased during the 20th century, as did the CO_2 concentration, and (2) climate models predict even greater increases in temperature in the 21st century as CO_2 emissions increase further.

Most of this climate modeling has been carried out on a global basis, but a few modeling studies were used to forecast changes in Arctic ice cover area to global warming induced by increased greenhouse gases. In this connection, coupled ocean–atmosphere general circulation models have usually been employed (e.g., Manabe and Wetherald, 1975; Vinnikov et al., 1999; Katzov, 2003; Johannessen et al., 2004). In addition to greenhouse gases, some of the models accounted for the influence of sulfate aerosol that forms in the troposphere as a response to warming and somewhat mitigates the role of greenhouse gases. The results of these model calculations show significant temperature increases due to predicted doubling of CO_2 in the 21st century. However, the models greatly amplify this temperature increase due to an assumed global increase in atmospheric humidity. However, the models have difficulty accounting for the dynamic effects of clouds, aerosols, regional variations in humidity, oceanic changes, variability of wind fields, and other factors. As a result, depending on the assumptions they make, the results vary over a wide range. For example, the international Coupled Model Intercomparison Project (CMIP) compared several of these models and showed that their results differ by several times. Izrael et al. (2001) noted that the least effective components of modern global climate models are parameterizations of sea ice and cloudiness and their feedback mechanisms. Thus, we believe that these model projections of future ice area fluctuations are unreliable.

When Vinnikov et al. (1999) analyzed results of a set of climate models, they found that the models forecast a decrease in the mean annual ice extent in the Northern Hemisphere of 1.2–1.6 million km^2 during the first half of the twenty-first century, followed by further acceleration of the decrease. Based on the same preconditions, a forecast of ice cover concentration during the summer period in 2081–2090 by Johannessen et al. (2004) indicates that a small area of ice with a concentration of 1–5 tenths will remain in the Arctic Basin by that time. These models are in line with the majority of climate models that predict a several-degree temperature rise in global average temperature, with even greater temperature increases in polar regions, due to a doubling of CO_2 concentration in the atmosphere later in the 21st century. However, Makshtas et al. (2007) show that reconstructed SAT values significantly overestimate the real temperature values measured at the "North Pole" drifting stations with an average systematic error in the summer months equal to $+1.2°C$. There appear to be large errors in the model estimates of cloudiness and air humidity, which subsequently leads to distortion of the heat flux values, and hence of the simulated sea ice thickness and concentration.

Most climate models have concentrated on predicting a future global average temperature rise due to a doubling of CO_2 in the 21st century compared to the pre-industrial level of about 280 ppm. Only a few of these models have attempted to analyze global temperature variability in the past. Of particular note is the fact that global temperatures dipped from about 1940 to 1978, while CO_2 emissions increased. Several papers tried to explain this with computer models including the negative effect of aerosols (e.g. Nagashima *et al.*, 2006 and Nozawa *et al.*, 2005). However, as Rapp (2008) pointed out: "these papers appear to raise more questions than they answer." Inconsistencies in the changes over time of anthropogenic carbon dioxide emissions to the atmosphere and anomalies of global surface air temperature are convincingly presented in Klyashtorin and Lyubushin (2003).

In general, although climate models are based on physics, they inevitably include a number of adjustable parameters that are fitted to past temperature changes. We are not aware of a single climate model based on fundamental physics without adjustable parameters that has been subjected to a rigorous test against actual climate data. Climate modelers appear to assume that the Earth's climate would continue without change, were it not for greenhouse gas emissions. They do not take into account the possibility that natural climate cycles are also acting independently of effects induced by buildup of greenhouse gas concentrations. As we have shown in Chapter 4, there is evidence for cyclic variability of Arctic climates. Furthermore, there is considerable evidence for past variability of global climate as expressed in the so-called Medieval Warm Period (900–1100) and the Little Ice Age (1600–1850). These fluctuations appear to be as great as the temperature rise of the 20th century, yet, there was no contribution of greenhouse gases to these climate changes.

A major challenge in climate modeling is to understand the range of natural fluctuations, and separate these from climate changes induced by human activity (greenhouse gas emissions, land clearing, irrigation, . . .). The models neglect natural fluctuations because they have no means of incorporating them, and put the entire blame for climate changes since the 19th century on human activity. As a result, they appear to project an extreme view of the future that seems unlikely to be reliable.

In Figure 5.1, the dynamics of global air temperature anomalies obtained from instrumental measurements over the last 140 years is compared with changes in world fuel consumption (WFC) (Makarov, 1998). The WFC curve shows an exponential increase, which doubles approximately every 30 years, increasing 25-fold since the middle of the nineteenth century. The global air temperature anomaly curve shows a positive trend of $+0.06°C/10$ years (Sonechkin *et al.*, 1997). At the same time, there are cyclic changes with periods of about 60 years. The correlation between these curves changes its sign every 30 years, varying from -0.88 (1940 1970) to $+0.94$ (1970 2000). Hence, there is no direct linear connection between WFC (which indirectly represents CO_2 concentration in the atmosphere) and global air temperature. The authors of this study therefore conclude that the WFC increase is not an obvious cause of the increase in global air temperature.

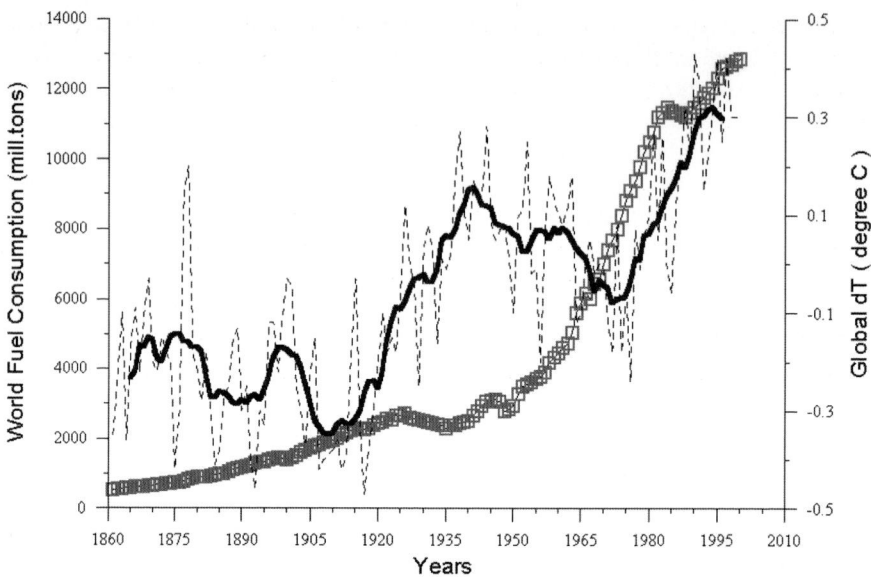

Figure 5.1. Comparative dynamics of the World Fuel Consumption (WFC) and Global Surface Air Temperature Anomaly (ΔT), 1861–2000. The thin dashed line represents annual ΔT, the bold line—its 13-year smoothing, and the line constructed from rectangles—WFC (in millions of tons of nominal fuel) (Klyashtorin and Lyubushin, 2003).

Divine and Dick (2006) refuted the idea that changes in sea ice extent in the Nordic Seas from the second half of the nineteenth century to the end of the twentieth century resulted from the superposition of a natural 60–80 year fluctuation and a long-term trend caused by the "greenhouse effect," because the latter was clearly pronounced in the first part of the indicated period when "any anthropogenic impact was still negligibly small."

Sorokhtin's adiabatic greenhouse-effect theory (Kuznetsov and Sorokhtin, 2000; Sorokhtin, 2001) is of particular interest in this connection. He convincingly critiqued the idea that anthropogenic emissions of greenhouse gases have a decisive influence on Earth's climate. His theory, based on simulations and observational data, suggests that the temperature distribution in the troposphere is determined by convection rather than by radiation processes. Sorokhtin formulated several consequences of his theory that contradict the results of the model simulations mentioned above. In his opinion, the main factors responsible for the state of Earth's climate are solar radiation, solar activity, and the composition, pressure, and heat capacity of the atmosphere. An increase or decrease in carbon dioxide in the atmosphere is not a cause but rather an effect of the temperature change because the solubility of this gas in water decreases with increasing water temperature. The same conclusion was drawn earlier by Monin and Shishkov (1992). Aleksandrov *et al.* (2004) conclude that the "observed Arctic Basin variations in air temperature are in many respects inconsistent with the presumed climate changes simulated by climate models as responses to the greenhouse effect" (p. 138–140).

In order for climate models to estimate climate change for any time interval, they must include the principal mechanisms leading to these changes. These models must describe and explain the observed state and variability of atmospheric circulation, air temperature, ocean circulation, sea ice movement, and many other parameters; unfortunately, these climate models are presently unable to do this sufficiently well. Therefore, we agree with Kondratyev (2004, p. 121) that the "observational data ... by no means contain a clear confirmation of the existence of anthropogenically determined global warming," while "the results of numerical climate modeling substantiating a hypothesis of greenhouse global warming and supposedly consistent with the observational data present no more than adjustment to the observational data." Many well-known scientists hold the same opinion (about 80 publications by such "unorthodox" authors are cited in Kondratyev (2004). Many scientists oppose the "greenhouse theory." Dobrovolsky (2000, 2002) summarizes the opinion of many alternative environmentalists worldwide that scientific data refute the existence of the greenhouse crisis. They are supported by dozens of prominent climatologists, whose studies are also reviewed by Dobrovolsky (2000, 2002) and Schulte (2008). However, the majority of climatologists favor the hypothesis of greenhouse global warming (Oreskes, 2004; Oreskes *et al.*, 2008). The problem is that polarized viewpoints seem to have hardened into belief systems, almost like religions, whereas there are insufficient data to be certain about causes of climate change. As Tom Sawyer said in Mark Twain's classic: "Making predictions is difficult, especially about the future."

While many climatologists are convinced of the decisive role of the accumulation of anthropogenic greenhouse gases in the atmosphere as the cause of a future catastrophic warming of the Earth with a significant decrease in the Arctic Ocean ice cover, this theory has the following generic weaknesses:

— There are large discrepancies in the results of simulations of climate change using coupled atmosphere–ocean models, which testifies to the uncertainties inherent in the models.
— These models are unable to simulate real historic climate changes.
— There have always been fluctuations in the Earth's climate that lie within the range of warming in the 20th century.
— The evidence of global warming (global warming, glacier retreat, sea surface warming, ...) began prior to large scale CO_2 emissions (~1850–1880) and the rate has not been changed much with increased CO_2 emissions over time.
— The temperature increase in the 20th century associated with a 100-ppm increase in CO_2 is much smaller than the temperature change associated with 100-ppm variations in ice age cycles.

In addition to these generic issues, the following inconsistencies occur in regard to specifically Arctic phenomena:

— Thickness changes in landfast ice of the Arctic Seas, where only thermodynamic processes are active, are small and considered to be unreliable.

— Significant "thinning" of drifting ice during the last few decades of the 20th century has no direct relevance to the increased concentration of greenhouse gases, but rather is explained by relatively short-term anomalies in ice-cover dynamics.

— Data from numerous ice-drift observations do not confirm the increased drift speed in the Arctic Basin from the middle to the end of the 20th century that is assumed by climate modelers: the average ice drift speed in the Arctic Basin for monthly and six-month periods in the periods both before 1975 (ice drift data available from manned stations and DARMS) and after 1975, through 2000 (ice drift data available from manned stations and IABP buoys), nearly coincides, while ice export to the Greenland Sea increases during cold epochs and decreases during warm epochs (see Figure 4.14).

— Analysis of sea level change in the Arctic Seas (Proshutinsky *et al.*, 2004) for 1954–1989 (0.189 cm/year) indicated that most of these changes can be explained by natural causes (steric, inverse barometer, and wind effects) while "the residual term of the sea level rise balance assessment, $0.048 \, \mathrm{cm \, yr^{-1}}$, may be due to an increase in the Arctic Ocean and global ocean mass associated with melting of ice caps and small glaciers and also with adjustments of the Greenland and Antarctic ice sheets to the observed climate change.

— Claims of a constant decrease in the multiyear ice in the Eurasian sector of the Arctic Basin during the second half of the twentieth century are not correct: the boundary of prevailing old ice exhibited from 1–2 years to multidecadal variations by the turn of the century was similar or even closer to the coast in comparison to that for 1950s for the seas of the eastern region.

5.2   THE INFLUENCE OF SOLAR ACTIVITY ON CLIMATE AND THE ICE COVER

Solar activity (SA) includes a complex of physical phenomena that take place on the Sun that lead to variations in solar emissions (electromagnetic and corpuscular). Various indicators are used for quantitative characterization of SA. These include sunspot indices, solar cycle duration, changing diameter of the Sun, geomagnetic indices, solar wind indices, etc. The most widely used is the Wolf number, which expresses a relative number of solar spots and their groupings on the visible solar disc. Changes in the Wolf number over time have allowed detection of their cyclicity ("Schwabe–Wolf's law"; Vitinsky, 1973; Rapp, 2008). The average duration of this cycle is presently 11.1 years, but this has varied widely over the past centuries. The cycles are numbered in a system known as Zurich numbering (Vitinsky, 1973; Rapp, 2008). Other indicators, discussed below, are also applied in studies of SA that characterize various aspects of solar activity and its influence on geophysical phenomena. Cycles detected for various indicators include those lasting 22 years and 80–90 years (Vitinsky, 1973).

Scientists' opinions on the role of SA in climate change on Earth differ significantly, from complete denial that it has any role (Monin, 1969) to attributing full determination and control (Bashkirtsev and Mashnich, 2004; Yegorov, 2004). The first to investigate the relationship between solar activity and sea ice extent was Vize (1944b,c). Comparing the sea ice extent of the Barents Sea using annual Wolf numbers, he found that the correlation coefficients characterizing this relationship have quite high values during some periods; however, the sign of the relationship changes from one period to another.

Maksimov (1954, 1955, 1970) and his students undertook major studies of solar activity's influence on the sea ice extent of the Arctic Seas. As early as 1954–1955, he proposed a "component-harmonic method of calculation and forecast of sea ice extent in some regions based on a periodogram analysis of a series of annual observational data." In addition, to "solar-based" 11-year and century-long fluctuations, the method took into account a 6-year cycle generated by the pole tide and a 19-year cycle connected with a long-period lunar declination tide. Further development of this method resulted in a "component-genetic" method: instead of using the results of a periodogram analysis, some component relationships were used to characterize an imaginary periodic part of the forecasted characteristic, expressed both by Wolf numbers and by specially calculated coefficients (Maksimov, 1970). A relationship between the total sea ice extent of the Arctic Seas and solar activity (the Wolf numbers) was claimed by Kovalev (1967). He found that the sign of the relationship changes from one period to another, and explained it by the fact that the inverse relationship of sea ice extent and SA is invoked when the 11-year fluctuations of sea ice extent of helio-physical origin are in a phase opposite to fluctuations of shorter, 6–8-year (geophysical) periods.

In subsequent years, several studies were published in which the Wolf numbers were used as an indicator of SA to make forecasts of sea ice extent of some Arctic Seas and their regions (Chaplygin and Yanes, 1968; Santsevich, 1970; Karklin, 1977; Karklin and Teitelbaum, 1987). Kupetsky (1969, 1974, 1977) describes a method for superimposing even and odd cycles of SA to forecast sea ice extent. In addition to the Wolf numbers, an index of geomagnetic planetary perturbation M, proposed by Ol' (1969), was also used. However, the statistical significance of such methods was questioned by Kovalev and Spichkin (1977).

Lassen and Friis-Christensen (1991) also noted the similarity between long-term variations in the Greenland Sea ice-cover area and SA expressed by Wolf numbers, and they pointed out the ambiguous relationship between SA and sea ice extent west and east of Greenland. The same study reveals a close correlation (-0.95) between the surface air temperature in the Northern Hemisphere (for 1861–1989) and the duration of the SA cycle: warming exhibits shorter cycles (about 10 years) and cooling longer cycles (about 11.5 years). Very early in the twentieth century Meinardus and Shott had noted opposite sea ice extent conditions west and east of Greenland determined by atmospheric circulation regimes (Alekseev *et al.*, 1998).

A study by Sleptsov-Shevlevich (1991) claimed an important role for a 22-year ("magnetic") SA cycle in interannual variations in sea ice extent of the Arctic Seas as well as other oceanographic, meteorological, and geophysical phenomena. This cycle can be correlated with the change in sign of the Sun's magnetic fields at the transition from one 11-year cycle of SA to another. This author concluded that a 2-year cycle of sea ice extent and a 6–7-year cycle of fluctuations in the location of the North Pole occur together within a 22-year cycle.

These ideas were further developed in studies by Latukhov and Sleptsov-Shevlevich (1995) and Sleptsov-Shevlevich and Zakharov (1996) who focused on "100-year" variations in sea ice extent that correlate with the magnetic perturbation index K_p, which is related to SA: when K_p increases, the sea ice extent of the sub-Atlantic Arctic Seas east of Greenland decreases while it increases in East-Canadian waters, and vice versa. The first of these studies also analyzes the causes for error in a super-long-range forecast of sea ice extent in the sub-Atlantic Arctic Seas by Maksimov (1955) and proposes a new forecast based on the updated duration of the 100-year cycle of solar activity: the epochal maximum sea ice extent in this region is expected around 2023.

Close relationships between the total number of large anomalies ($\geq 1.2\sigma$) in Arctic Seas sea ice extent (regardless of the sign of the anomalies) during 11-year SA cycles and the sum of the Wolf numbers in the corresponding cycles were revealed by Karklin and Kovalev (1994; Figure 5.2). Averaging of the Wolf numbers for each 11-year SA cycle, performed in this study as well as in some other studies, excludes an analysis of the dependence of ice conditions on SA fluctuations within the 11-year solar cycles. The relationship depicted in Figure 5.2 reflects the influence of longer SA variations.

Shirochkov and Makarova (1998) outlined a new approach, similar to that for other Earth climate characteristics, for studying the relationship between long-period changes in SA and sea ice extent in the Arctic Seas. The solar wind dynamic pressure (P_{sw}), which depends on the density of particle fluxes from the Sun to the Earth and

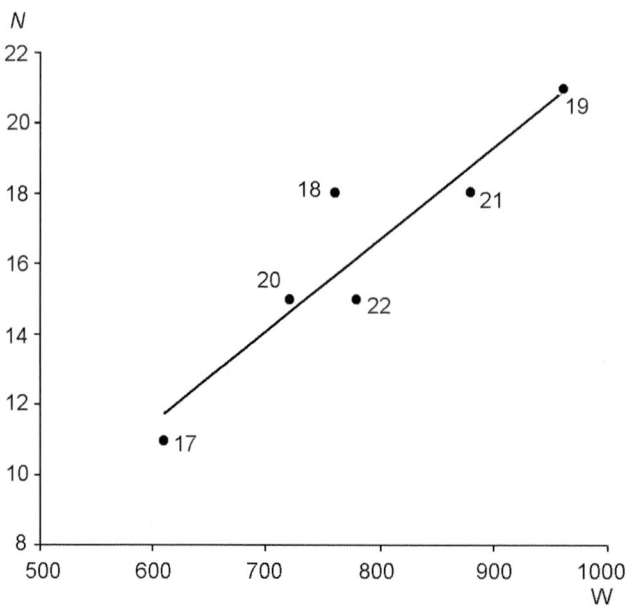

Figure 5.2. Relationship between the number of large sea ice extent anomalies (N) in the Arctic Seas in August–September to the total value of Wolf numbers (W) in 11-year solar cycles. Figures near the points indicate numbers of 11-year solar cycles using Zurich numbering.

their speeds, was used as an indicator of SA. The value of P_{sw} is measured by special satellite instruments. An analysis of the relationship between the P_{sw} index and the ice-cover area in the Greenland, Barents, and Kara seas, and the sea ice extent of East Canadian waters for April–July, suggested a strong inverse relationship for the Greenland and Barents Seas: increased solar wind pressure is accompanied by decreased sea ice extent (the correlation coefficients are -0.77 and -0.64, respectively), whereas for East Canadian waters (mainly Baffin Bay), a weak direct correlation (0.30) is noted. This appears to confirm the conclusions reached by comparing sea ice extent of the study areas with SA expressed by Wolf numbers, considering that there is a significant although unstable negative correlation between the SA indicators used: from -0.60 for the period 1964–1980 to -0.33 for the period 1980–1996 (Shirochkov and Makarova, 1998).

This far-from-complete review of studies concerning the relationship between sea ice extent and SA indicates the presence of possible little-studied natural mechanisms that might affect these putative relationships.

At present, there is a broadly held opinion that variations in solar activity are primarily expressed as changes in atmospheric pressure fields and atmospheric circulation that result in anomalies of other hydrometeorological elements (Karklin, 1973). Atmospheric pressure changes at sea level with periods of about 11 and 22 years have received the most attention.

Studies of air pressure field changes caused by solar activity (Maksimov, 1970; Sleptsov-Shevlevich et al., 1991) reveal standing waves in Earth's atmosphere, with periods corresponding to known cycles of solar activity. These include the 11-year sunspot cycle (its duration varies from 8 to 17 years), and a 22-year cycle (varying from 18–28 years), which is related to two things: 1) the discovery by Hale that the sign of solar spots' magnetic polarity changes from one 11-year cycle to another (Vitinsky, 1973), and 2) features of changes in magnetic perturbations that were revealed by Ol' (1969).

The first of these waves is characterized by a stable positive relationship between atmospheric pressure and solar activity (the Wolf numbers) in the high latitudes of the Northern Hemisphere and by a negative relationship in temperate latitudes. The position of the wave's nodal line changes from cycle to cycle, which results in an extensive zone of unstable solar–atmospheric relations encircling the high-latitude area of stable direct relations with the boundary in the North European Basin at approximately $70°\text{N}$ (Maksimov and Sleptsov-Shevlevich, 1963; Gasyukov and Smirnov, 1967; Karklin, 1978).

Table 5.1 indicates the significance of solar activity in the variability of atmospheric pressure fields over the Arctic Ocean using the average values of large-scale wind-field vorticity in years of increased and decreased SA expressed by Wolf numbers. The characteristics that indicate the intensity of cyclonic (anticyclonic) circulation were calculated for two regions of the Arctic Ocean: the North European (J_1) and Arctic (J_2) basins, averaged for the winter months (October–March). For these calculations, equations expressing the Laplacian of atmospheric pressure distribution similar to Equation 4.11 were applied. To calculate the J_2 value, a triangle replaced the squares used for calculating J_0 and J_1 (see Figure 4.22).

Table 5.1. Wolf numbers ($\sum W$), cyclonicity indices in the western (J_1) and eastern (J_2) regions of the Arctic Ocean, and total sea ice extent (thousand km^2) in the western (L_1) and eastern (L_2) regions during cycles of increased and decreased solar activity

Period, years	$\sum W$	J_1	J_2	L_1	L_2
1937–1941	429	35	−63	735	1031
1942–1946	183	59	−67	792	1003
1947–1951	441	51	−73	864	1041
1952–1956	229	85	−3	689	1001
1957–1961	700	69	−60	795	940
1962–1966	120	62	−36	1163	1157
1967–1971	465	87	−65	1210	941
1972–1976	170	100	−38	889	1087
1977–1981	498	62	−68	1057	1022
1982–1986	180	80	−44	829	1180
1987–1991	471	79	−36	953	821
1992–1996	192	105	−20	742	941

Table 5.1 provides Wolf number sums for 5-year time intervals, characterizing the periods of increased and decreased SA (cycles 17–22). It also shows the values of J_1 and J_2 averaged for October–March, expressing the intensity of cyclonic (anti-cyclonic) circulation in the indicated regions (Figure 4.22), and the sea ice extent values averaged for the same 5-year periods for the western (L_1) and eastern (L_2) regions. Based on these data, corresponding average values were calculated to characterize the periods of increased and decreased SA during 17–22-year cycles (Table 5.2).

As Table 5.2 shows, at an almost triple (on average) change in the Wolf numbers during the 11-year cycle—from a 5-year period of increased SA (501) to a 5-year period of decreased SA (179)—the intensity of the Icelandic cyclonic circulation slightly increases on average (from 64 to 82 units), and the intensity of the anti-cyclonic circulation in the Arctic High drops significantly (on average from −61 to −35 units). That is, the cyclonicity increases in both regions. These results match the character of the 11-year fluctuations in the baric field discussed above: with an increasing Wolf number, the atmospheric pressure in high latitudes increases, and

Table 5.2. Wolf number ($\sum W$) averages for periods of increased and decreased solar activity, cyclonicity indices for the western (J_1) and eastern (J_2) regions of the Arctic Ocean, and total sea ice extent (thousand km^2) in the western (L_1) and eastern (L_2) Eurasian Arctic regions

Periods	$\sum W$	J_1	J_2	L_1	L_2
Increased solar activity	501	64	−61	936	966
Decreased solar activity	179	82	−35	851	1061
Difference	*322*	*−18*	*−26*	*85*	*−95*

with a decreasing Wolf number, it decreases (Karklin, 1978). Increased cyclonicity is accompanied by decreased sea ice extent in the western region and increased sea ice extent in the eastern region. Increases in anticyclonicity have the opposite effect on the ice cover area in both regions. It is important to note that the influence of solar activity on the baric field is more pronounced in the Arctic Basin than in the North European Basin. This pattern can be probably be explained by the phenomenon noted above that a nodal line dividing the areas of the standing solar-determined baric wave with a different sign of corresponding anomalies often passes across the North-European Basin.

In analyzing the influence of the 11-year SA cycle on atmospheric circulation, it is important to remember that satellite data available since 1978 show that the difference between maximum and minimum solar radiation in the 11-year cycle is only 2 W/m^2 (0.15% of the average solar constant value) (Bashkirtsev and Mashnich, 2004; Rapp, 2008). Therefore, explaining the manifestation of this cycle in the Earth's atmosphere requires accounting for the corresponding variations in UV radiation, particle fluxes, galactic rays, and the presence of trigger mechanisms that could cause energetically insignificant variations in incoming solar radiation to result in significant weather changes in the Arctic.

The character of the spatial distribution of the relationship between atmospheric pressure and solar activity in a 22-year wave is similar to an 11-year wave both in the location of the loops at high and temperate latitudes and in the location of the nodal line (zones of unstable relations). During even numbered 11-year cycles (using Zurich numbering), atmospheric pressure decreases in the near-pole region and increases at temperate latitudes. On the contrary, during odd numbered cycles, the atmospheric pressure increases at high latitudes and decreases at temperate latitudes. The nodal line of this wave passes between 55°N and 60°N (Maksimov and Sleptsov-Shevlevich, 1971; Ol' and Sleptsov-Shevlevich, 1972; Karklin, 1973). Similar fluctuations in atmospheric pressure were also observed in the Southern Hemisphere (Maksimov and Sleptsov-Shevlevich, 1963).

Differences in atmospheric circulation during odd and even SA cycles are convincingly reflected in average NAO index values (Gudkovich et al., 2004). The average twentieth century NAO anomaly was predominantly positive in the even cycles and negative in the odd cycles. These results suggest a high probability of intensified zonal transports in the atmosphere of temperate latitudes during even SA cycles and their weakening during odd cycles, which is consistent with the behavior of the 22-year wave.

According to available estimates (Karklin, 1978), the contribution of solar-induced fluctuations to the variability of atmospheric pressure is 10 to 30% for both the 11- and 22-year waves, which influences the general circulation of the atmosphere, especially the intensity of west-to-east air flow at temperate and high latitudes. This is confirmed by investigations of the relationship between SA and recurrence of the main atmospheric circulation forms (Girs, 1960) and the number of elementary synoptic processes during a year (Dmitriyev, 1994). Some studies found that the location of atmospheric action centers and their development changes with the 11-year SA rhythm (Abramov, 1967; Karklin, 1975).

The relationship between SA and Arctic baric fields influences a complex of factors that determine the long-term variability of the ice state in the Arctic Seas. One of these factors is ice export from the Arctic Basin to the Greenland Sea. A monthly comparison of ice exported through Fram Strait from 1946 to 1999 was carried out using the method developed by Gudkovich and Nikolayeva (1963) and averaged for each of the last five 11-year cycles of SA, using the Wolf-number average for each cycle. It indicated that the ice export increases with an increase in the Wolf number average during odd SA cycles and decreases during even cycles (Figure 5.3). This pattern helps to explain the dependence of the ice cover state of the Arctic Seas on SA.

S, 1000 km^2/month

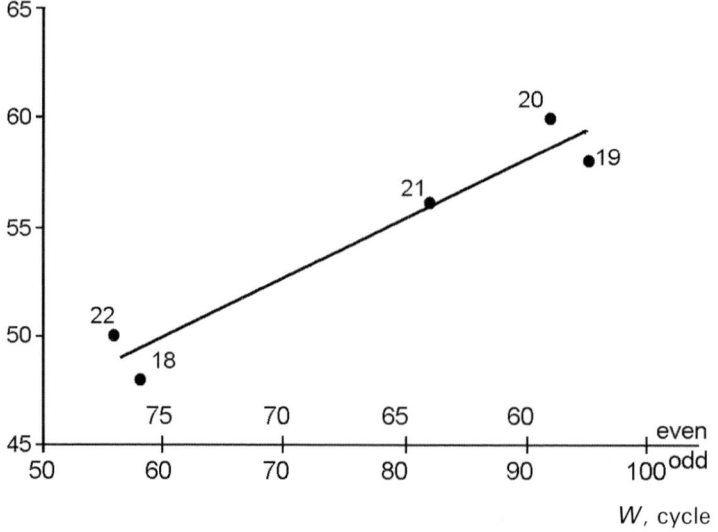

Figure 5.3. Dependence of monthly ice export area (S) from the Arctic Basin to the Greenland Sea on the Wolf number average (W_c) for a cycle during the odd (19, 21) and even (18, 20, 22) cycles of solar activity.

Bashkirtsev and Mashnich (2004) provide evidence of relationships between SA and various hydrometeorological indicators in different regions. Zherebtsov and Kovalenko (2001) note a close correlation (0.97) between averaged 11-year solar cycle Wolf numbers and surface air temperatures in the Baikal area.

Studies by Reid (2000) and Makarov and Tlatov (2000) conclude that variations in surface air temperature over the oceans are similar to Wolf number variations. Multiyear fluctuations of global surface air temperature (GSAT) exhibit periodicity similar to solar activity: the Schwabe cycle (11 years), the Hale cycle (22 years), and the Fritz cycle (about 60 years) which are also evident in the Sun's large-scale magnetic field and in the aurora borealis (Bashkirtsev and Mashnich, 2004). Diagrams shown by Bashkirtsev and Mashnich (2004), based on the SA data obtained by Nagovitsyn et al. (Nagovitsyn et al., 2004; Nagovitsyn, 2007), illustrate the similarity of smoothed variations of SA and GSAT for 1100–2000 and extrapolated to 2300. SA variations for 1611–2005 in a form of yearly sunspot numbers are reproduced in Figure 5.4. Smoothed curves in this figure suggest cycles lasting about 200 years, which can explain the intra-secular trends considered above. The figure also reflects Earth's coldest period in the last millennium, coinciding with the known Maunder

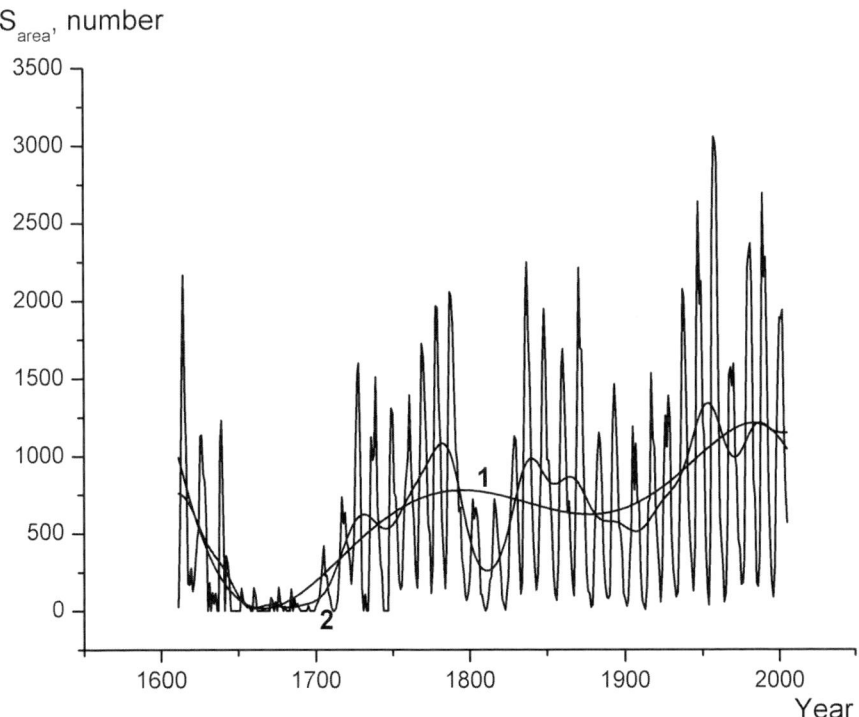

Figure 5.4. Observed and simulated yearly sunspot areas (Greenwich general system) for 1611–2005 (Nagovitsyn et al., 2004; Nagovitsyn, 2007). (1) Approximation by a polynomial to the sixth power. (2) FFT-filter by 11 points (years).

minimum in SA variations (1645–1715). The same article by Bashkirtsev and Mashnich (2004, p. 136) also provides an important scheme for interpreting the results set forth in Section 5.4: "the largest GSAT occurs during Hale and Fritz cycle synchronism: 1880, 1940, 2000 and 2060." This provides additional evidence for a 60-year cycle in Earth's climate fluctuations.

When considering the relationship of variations in the baric field and SA both lasting more than a century, an increase in the Wolf number corresponds to intensified cyclonic activity over the Arctic Basin. For example, Gudkovich *et al.* (2005) found that the average rate of increase in the Wolf number in the twentieth century was about 50 units/100 years. In the opinion of Gribbin and Lamb (1978), the assumption that such long-term relationships exist strongly suggests the importance of investigating the processes occurring in the atmosphere of our planet.

The discussion above suggests that large-scale changes in atmospheric circulation can be caused by SA at some time scales. However, the physical mechanisms for interaction between solar processes and Earth's troposphere are not yet resolved. Studies devoted to this issue include Maksimov (1970a), Mustel (1974), Vitinsky *et al.* (1976), German and Goldberg (1981), Krymsky (1994), Kondratyev and Nikolsky (1995), as well as many others. Proposed mechanisms range from the possible influence of SA on gravity and the solar constant to hypotheses regarding the impact of chemical, condensation, and electrical processes induced by anomalies of wave or particle solar energy fluxes. After satellite and radiosonde observations showed that solar wind causes heating and expansion of the upper layer of Earth's atmosphere, questions arose regarding the mechanisms for energy transfer from the stratosphere and magnetosphere to the troposphere.

Based on the law of conservation of momentum in the atmosphere, Sytinsky (1987) concluded that heating of the upper layers of the atmosphere by a solar particle flux causes the atmosphere's moment of inertia to increase, resulting in a decrease in the angular rotation speed of the atmosphere relative to the Earth, and the atmospheric pressure distribution gradually adapts to this change. A decrease in particle flux, whose energy is largely absorbed near the poles, has the opposite effect. These effects are also influenced by the direction of interplanetary magnetic fields, which change at the transitions between odd and even solar cycles. According to Bothner and Schwenn (1998), the magnetic fields of solar emissions during odd cycles are primarily directed opposite to Earth's magnetic field, which allows solar particles to enter Earth's atmosphere and leads to an increase in GSAT.

Theoretical studies by Krymsky (1994) concluded that the solar wind transfers not only mass, energy, and impulse to the magnetosphere, but also the moment of impulse, which is then transferred to the atmosphere and to the Earth as a result of turbulent viscosity. In this way, solar activity causes variations in the super-rotation of the atmosphere, in zonal circulation intensity, in atmospheric pressure distribution, and even in Earth's angular speed. The angular speed also directly depends on the rotation moment and corresponding torsional oscillations excited by the solar wind in the electromagnetically connected mantle and core of the Earth. Krymsky (1994) categorically rejects the opinion of some scientists regarding the transfer of the moment of impulse to the atmosphere from the Earth, because this would require "Earth, at insignificant angular speed changes, to impart to the atmosphere a rotation with a speed one million times greater" (p. 42). The proposed theory purports to explain, at least qualitatively, many phenomena in the atmosphere and solid Earth that are influenced by SA: cyclic changes in atmospheric circulation, angular speed of Earth's rotation, nutation of its axis, etc.

A different approach is used by Sleptsov-Shevlevich (1991) who assumes that changes in the angular speed of Earth's rotation due to solar wind lead to corresponding gravity changes as a sum of gravity and centrifugal force, causing cyclic "deformations of all Earth's shells," i.e., redistribution of the water and air masses, climate changes, etc. A decrease in the Earth's angular speed leads to displacement of air masses toward high latitudes, and its increase displaces the air masses toward mid latitudes, which influences the processes of cyclogenesis, prevailing trajectories of cyclones, etc. In our opinion, one of many weak points in this hypothesis is the fact that the changes in the atmosphere can occur only after Earth's angular speed variations are transferred to its air shell. As shown above, real mechanisms for this are absent.

Shirochkov and Makarova (1998) conclude that changes in solar wind pressure significantly affect the thermal regime of the middle atmosphere and even the state of the tropopause in Earth's polar regions. When solar wind pressure causes heating of the lower stratosphere, the tropopause "thins," which has some specific climatic consequences. These authors suggest a version of global electric circulation, which they have improved, as a possible mechanism for this relationship. An important element of this mechanism is a "giant spherical capacitor" with a magnetopause as its external plate and Earth's surface as its internal plate. The energy accumulated by this capacitor increases with a decrease in the distance between the plates, which is subject to the influence of solar wind pressure. The local character of the influence of this parameter on hydrometeorological processes can be explained by the differences in electrical conductivity of the underlying surface of the water (for example, in the Greenland Sea and Davis Strait).

The latter assumption is, in our opinion, the weakest point in this interesting hypothesis. It appears that the known phenomenon of the "tropopause funnel" (a significant decrease in the tropopause above the deep cyclone of low mobility; Khromov and Mamontova, 1974) should be examined as a possible physical mechanism for the relationship between the state of the tropopause height and processes in the troposphere. It is possible that the changes in the tropopause structure caused by solar wind pressure variations lead to significant changes in cyclogenesis in some specific regions, for example, at atmosphere action centers. If the increased solar wind pressure is accompanied by deepening of the Icelandic Low and its decrease with depression, then the ambiguity of the relationship of P_{sw} to sea ice extent in the Greenland and Barents seas on the one hand, and the ice cover area in Davis Strait, on the other hand, becomes clear. Deepening of the depression contributes to increased heat advection to the former regions and cold advection to the latter, and vice versa. The facts confirming this hypothesis are presented below.

To reveal the relationship between solar wind pressure and the Icelandic Low, the anomalies of atmospheric pressure at sea level for October–March, averaged by five points in square 1 of Figure 4.22 (1965–1995), were compared with the anomalies P_{sw} for the same period. In 70% of the cases, the anomalies in the two categories had opposite signs. This means that an increase in solar wind is typically accompanied by a decrease in atmospheric pressure in the area of the Icelandic Low, and, on the contrary, weakening solar wind results in an increase in atmospheric pressure. This supports the hypothesis that the solar wind impacts the baric field. There are grounds to suppose that the Arctic Oscillation (AO) phenomenon considered in section 4.2 reflects the influence of SA on Earth's baric field: the distribution of atmospheric pressure anomalies in the AO phases shown in Figure 4.8 closely matches the changes in atmospheric pressure fields at high and temperate latitudes during corresponding SA cycles (Karklin, 1973, 1978). The fact that both phenomena include cycles lasting about 10 and 20 years also supports an AO/SA relationship.

The 10-year cycle, the physical mechanism that causes corresponding changes in atmospheric pressure fields, is probably connected with the Wolf cycle of solar activity, which allows intrusion of charged solar wind particles into the upper layers of Earth's atmosphere at high latitudes. This would appear to cause an experimentally determined effect (Sytinsky, 1987) of atmospheric heating and expansion in the zone of particle intrusion, the corresponding appearance of horizontal atmospheric pressure gradients, and redistribution of air masses, which can trigger convection processes in the troposphere.

The "20-year" cycle of baric field oscillations over the Arctic Ocean and the sea ice extent of its seas, which are components of the AO, can also be related to the known 22-year fluctuation in solar activity (Hale's law) evident in solar magnetic field sign changes. Unfortunately, there are no convincing hypotheses on the possible mechanisms for such relationships. In some studies (Jose, 1965; Vasilieva, 1997; Vasilieva et al., 2002), processes occurring in the interior of the sun that cause magnetic field changes, fluctuations in the solar diameter, and other phenomena are explained by solar dissipative processes being determined by the distance between the center of the sun and the mass center of the solar system. The period of these fluctuations is close to the period of synodic revolution of Jupiter and Saturn (19.86 years).

In Raspopov et al. (2004), 20–25-year cyclic climate changes were revealed on the basis of a dendrochronological analysis of extensive data available on the northern forests of Arctic Eurasia for the period from 1458 to 1975. Their analysis of fluctuations in solar activity, based on measurements of radionuclide (^{14}C) concentrations in annual tree-trunk rings, identified the cyclic climate changes and solar activity under consideration here. Raspopov et al. (2004) conclude that as a result of nonlinear impact, the influence of solar activity and related cosmic ray fluxes can increase significantly (three- to fivefold) the inherent internal oscillations of the atmosphere–ocean–continent system. This may express the nature of double-ten-year global climate fluctuations. Potential mechanisms for such fluctuations are considered below.

5.3 POSSIBLE INFLUENCE OF SELF-OSCILLATIONS IN THE OCEAN–ICE–ATMOSPHERE SYSTEM

A number of scientists assume that some specific components of modern climate change are, in the words of Shuleikin (1953), "within our planet itself, within its liquid and gaseous (and maybe partly within its solid) shells." As early as the 1930s, Shuleikin began to assume that the "ocean–atmosphere–land system is a self-oscillating system."

The self-oscillations of some systems are known to differ in principle from other oscillating processes in that no periodic external forcing is required for their occurrence; they are determined by the properties of the system itself. Self-oscillations require a constant energy source and mechanisms to regulate the input of this energy to the oscillating system, as well as negative feedbacks that tend to return the system to equilibrium.

Shuleikin's self-oscillation scheme concerned the ocean-ice system. Later, it was extended to include the atmosphere because its interaction with the ocean is a significant cause of climate change. In two proposed self-oscillation schemes, the feedback mechanisms are driven by water stratification (Nikiforov and Shpaikher, 1980) and fractures resulting from ice-cover dynamics in the Arctic Basin (Alekseev, 1976) that influence heat exchange between the ocean and the atmosphere, which is known to cause changes in atmospheric circulation. Nikiforov (2006) proposes a theoretical basis for self-oscillations in the Arctic Ocean system: a "chain of mechanisms responsible for the occurrence of the oscillating regime in a definite frequency band" (p. 107). In this scheme, oscillations with periods of 4–8 years occur in the horizontal plane of interaction between the Arctic and the North European Basins of the Arctic Ocean, while interaction in the vertical plane induces low-frequency oscillations with periods of 20–30 years.

Zakharov (1996) devotes a great deal of attention to self-oscillation in the ocean–ice–atmosphere system as the most probable driver of natural processes in the Arctic. This author considers the ice cover an active climatologic factor determining, in particular, the intensity of the Arctic High and the southerly displacement of the Arctic atmospheric front and the related belt of cyclonic activity. Zakharov (1977) also identifies the layer of low-salinity surface water underlying the ice as the main control on ice cover area (at least in winter). Expansion of this low-salinity water is regulated by the freshwater balance of the Arctic Ocean (inflow of freshwater, its outflow to the Atlantic, and the excess of atmospheric precipitation over evaporation).

Based on these cause–effect relationships, Zakharov suggests a consistent conceptual scheme of self-oscillations in the atmosphere-ocean-ice system that explains current climate change. The following sequence represents this scheme: positive freshwater budget in the Arctic Ocean → increasing volume and area of surface Arctic water → ice cover expansion and atmospheric cooling → a southerly shift of the Arctic climatic front and the precipitation belt → decreased freshwater inflow to the Arctic Ocean → a negative freshwater budget in the Arctic Ocean →

decreased volume and area of surface Arctic water and ice expansion, and so forth, reversing the order of these phenomena.

There is, however, one weakness in this scheme. Cooling and expansion of the Arctic High, in some cases, actually leads to the southward shift of the Arctic High and the precipitation belt. But this should result in an increase in river runoff to the Arctic Ocean, rather than its decrease, as precipitation in the relevant river basins increases. There is much less excess of precipitation over evaporation at high latitudes than at more temperate latitudes (Anon. (G)). If this is the case, then instead of a negative feedback, there is a positive feedback (cooling → increased runoff, warming → decreased runoff), which would not result in self-oscillation. This is confirmed not only by the precipitation decrease in Eastern Europe during the first Arctic warming in the 1930s and the decrease in Caspian Sea level at the same time but also by an analysis of the relationship of Kara Sea ice conditions to river runoff and air temperature by Gudkovich *et al.* (1981). Their study of 40 years of observations made between 1936 and 1975 indicates that when the relationship between sea ice extent and runoff from the Ob' and Yenisey rivers was reliable, the air temperature decrease and increased runoff operated in tandem, and vice versa (the coherence function is 0.66 to 1.00). However, in the other cases, expansion of the Arctic High signaled its merging with the Siberian High, thus hindering west-to-east air circulation over the Asian continent. Siberian river basin precipitation and runoff to the Arctic Ocean are dependent on the intensity of this air transport (see section 4.7).

During the periods of Arctic warming induced by 60-year cyclic fluctuations of the climate system, as noted above, increased precipitation in temperate latitudes of the Eurasian continent and corresponding increase in river runoff were connected with intensified cyclonic activity over the Arctic Basin, a weakened Arctic High, and progressive increase in atmospheric zonal flows. There was a simultaneous decrease in sea ice extent of the Arctic Ocean seas along with a synchronous positive trend in river runoff, which also does not support the discussed self-oscillation scheme.

Another disadvantage of Zakharov's (1996, 1997) concept is that the author considers only a horizontal advective mechanism for the fluctuation in low salinity surface water distribution. Meanwhile, as Section 4.6 shows, vertical circulation driven by baric-field vorticity (salinification of surface water during strong cyclonic activity and freshening at its weakening) is also important.

The large-scale interaction of the ocean and the atmosphere is most pronounced in the so-called energy-active zones such as the Norwegian energy-active zone (NEAZO) in the North European Basin, where relatively warm and saline Atlantic Ocean waters meet cold and low salinity water exported from the Arctic Basin. The atmosphere–ocean interaction is regulated here by the contrasts of the underlying surface temperature, on which the intensity of cyclogenesis depends. Baric field vorticity influences vertical circulation of the water and its stratification, which affects the process of convection that brings heat from deeper layers to the surface of the ocean. Gudkovich and Kovalev (2002) express this interdependent chain of self-oscillations and the time lags between them as:

$$\ldots L^+(3) \to \Delta P^+(2) \to S^+(6) \to L^-(3) \to \Delta P^-(2) \to S^-(6) \to L^+ \ldots,$$

where L, ΔP, and S denote the sea ice extent, mean annual vorticity of the wind field, and surface water layer salinity, respectively; signs (+) and (−) express the maxima and minima of the indicated values, and figures in brackets show the corresponding average values of lags (in years). The period of these self-oscillations is, on average, 22 years. This is a typical series of self-oscillations: a 10-year increase in sea ice extent results in its decrease, and vice versa (a negative feedback). The components of this process include: dependence of the intensity of cyclonic activity on sea ice extent and the related temperature gradients of the water and the air, dependence of surface-layer salinification on cyclonic activity, and inverse dependence of sea ice extent on surface water salinity and air temperature in the region. The time lags between these processes are probably determined by their "inertia", because changes in some characteristics of the ocean or the ice cover "can result from prolonged accumulation of stochastic forcing by the atmosphere" (Alekseev, 1995, p. 195).

The prevailing period of about 20 years suggests the possible influence of solar activity whose interannual variations exhibit a pronounced 22-year (Hale) cycle. Further research is needed on two processes with similar time cycles: the solar activity-influenced formation of baric anomalies (see section 5.2) and the lunar long-period tide with a cycle of about 19 years (e.g., Maksimov, 1970; Sleptsov-Shevlevich, 1991). These factors may have a stabilizing influence on the cyclic processes in the ocean–ice cover–atmosphere system.

A hypothesis set forth by Dukhovskoy *et al.* (2004) proposes another example of self-oscillation in this system. This hypothesis attributes the fluctuations in decadal-scale /NAO indices to self-oscillation that results from heat and freshwater exchange between the Arctic Basin and the Nordic Seas. According to this hypothesis, during periods of weak interaction between these regions, low salinity surface water accumulates in the Beaufort Gyre area and causes the sea level to rise. At this time, a freshwater deficit in the Nordic Seas reduces the ocean vertical stratification leading to an increase of heat flux from the ocean to atmosphere, which causes intensified cyclonic activity in the atmosphere and decreased sea level. The growing sea level gradient between the two regions then leads to strong interaction as freshwater flows from the Arctic Basin to the North European Basin, and warm Atlantic water flows northward. The Arctic High then weakens, the air temperature increases, and both the sea level gradient between the regions and freshwater runoff decrease. As a result, the system gradually returns to a state of weak interaction. The model developed by Dukhovskoy *et al.* (2004) showed the period of this self-oscillation to be 10–12 years.

A weak point in this hypothesis is the absence in nature of a critical level gradient value at which the convergence of full flows in the Ekman layer of anticyclonic circulation attains an opposite sign that is inherent in the divergence conditions. After all, the accumulation of water with decreased density in the anticyclonic circulations and a corresponding sea level increase are restricted by the vertical circulation (downwelling and water outflow at depth). This is illustrated by such global ocean currents as the Gulf Stream and the Kuroshio, where baroclinic current speeds and sea level gradients are two orders of magnitude greater than those typical of the Arctic Ocean. No data suggest that vertical cross-circulation in such currents attains an opposite direction.

5.4   SOLAR SYSTEM DISYMMETRY AND ITS INFLUENCE ON SOLAR ENERGY FLUX TO THE EARTH

The causes of the major "60 year" cycle of ice extent fluctuations and corresponding hydrometeorological characteristics have recently become more clear (Gudkovich *et al.*, 2005). The global character of these fluctuations points to their possible relationship to astronomical factors, including the location of the solar system's center of mass. Kovalenko *et al.* (1987) came closest to solving this problem; they called the vector connecting the centers of mass of the Sun and the solar system, the "dissymmetry of the Sun." Their study and other subsequent studies suggest that the main factors in climate change include the influence of "dissymmetry" on solar activity, integral flux of solar radiation, and other factors (Zavalishin and Vinogradova, 1990; Vasilieva, 1997; Vasilieva *et al.*, 2002; Baidal, 2001). However, these studies did not resolve the problem considered here regarding a 60-year climate change cycle.

Studies by Monin (2000) and Kurazhov *et al.* (2004) that focus on solar dissymmetry influenced by two of the largest planets in the solar system, Jupiter and Saturn, are of special interest. These studies relate the 60-year climate cycle to the revolution of the solar system around a common center of mass. Monin (2000), who proposed that "in order to determine the location of the center of inertia, it is sufficient to consider three bodies: Jupiter, Saturn and the Sun," computed the solar orbit period of motion to be about 60 years.

For our calculations, we took into account the masses of the Sun, Jupiter, and Saturn; the average distances between the planets and the Sun; and their periods of revolution (using rounded values of 12 years for Jupiter and 30 years for Saturn) to show that the distance between the center of the Sun and the solar system's center of mass changes, with a range of $0.34-1.15 \cdot 10^6$ km.

Figure 5.5 locates Jupiter and Saturn in 10-year time intervals as they revolve around the solar system's center of mass (also see Figure 5.6 for additional explanation). When both planets are located to one side of the center of mass, the distance from it to the center of the Sun is maximal (years 0, 20, 40, and 60 of the conventional time scale). At intermediate points (years 10, 30, 50, and 70), the planets are located on different sides of the center of mass, so the distance to it from the center of the Sun is minimal, and the center of mass shifts toward Jupiter from the center of the Sun. Hence, the period of such variations comprises about 20 years.

The time (t, years) necessary for forming the first and the second types of the location of planets can be derived from the following equations:

$$360t\left(\frac{1}{12} - \frac{1}{30}\right) = \varphi_0 \pm 360n \quad \text{or} \quad 18t = \varphi_0 + 360n \qquad (5.1)$$

and

$$360t\left(\frac{1}{12} - \frac{1}{30}\right) = \varphi_0 + 180 \pm 360n \quad \text{or} \quad 18t = \varphi_0 + 180 \pm 360n, \qquad (5.2)$$

where $n = 1, 2, 3, \ldots$ is an integer number.

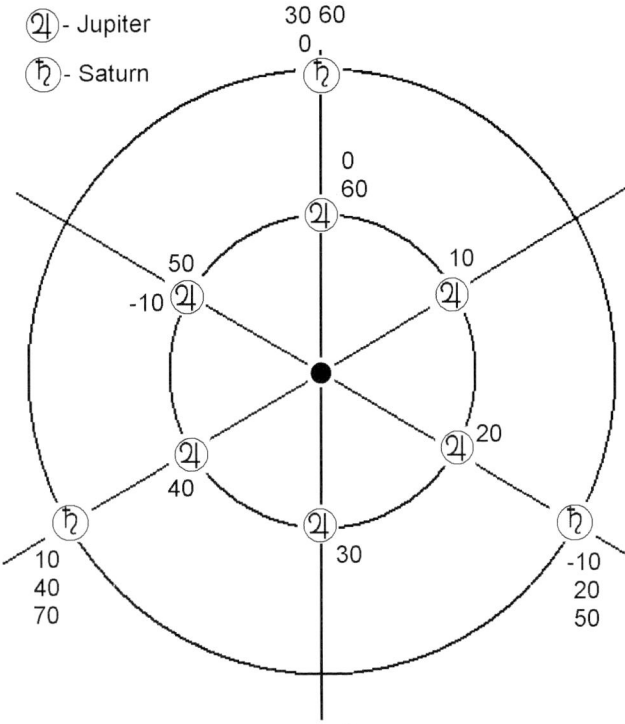

Figure 5.5. Scheme showing the locations of Jupiter and Saturn at different points in time (figures indicate the years from the initial moment $t = 0$) as they revolve around the center of mass (●) of the Sun–Jupiter–Saturn system (an arrow denotes the direction of motion of the planets).

From the values derived for t, the corresponding turning angles φ relative to the initial phase φ_0 can be determined from the expressions $\varphi = 360t/12$ (for Jupiter) and $\varphi = 360t/30$ (for Saturn).

Because of Earth's elliptical orbit, the average distance from the Earth to the center of the Sun changes during the year from approximately $147 \cdot 10^6$ km (in early January) to $152 \cdot 10^6$ km (in early July). According to Khromov and Mamontova (1974), changes in the extra-atmospheric intensity of solar radiation fall within approximately $\pm 50\,\mathrm{W/cm^2}$ due to Earth's annual motion. It is important to note that the location of the large orbit axis (line of apses) changes insignificantly ($6°$ for 2000 years; Ryabov, 1988).

Changes over time in the distance between the centers of Earth and Sun for the moments of Earth's perihelion and aphelion can be calculated by the equations:

$$CP = \sqrt{(L_p + \varepsilon \cos \varphi)^2 + (\varepsilon \sin \varphi)^2} \tag{5.3}$$

and

$$CA = \sqrt{(L_a - \varepsilon \cos \varphi)^2 + (\varepsilon \sin \varphi)^2}, \tag{5.4}$$

where CP and CA are the distances between the centers of the Earth and the Sun at perihelion and aphelion; L_p and L_a are the distances between the Earth and the solar

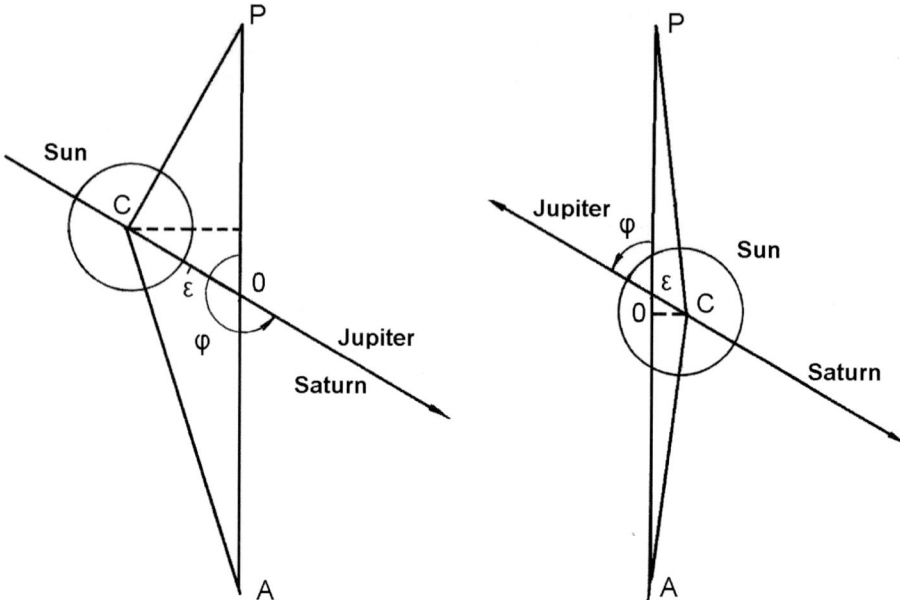

Figure 5.6. Calculation diagrams for the distances between the Earth and the Sun for two types of locations for Jupiter and Saturn relative to the Sun (C = center of the Sun; O = mass center; P and A = locations of the Earth at perihelion and aphelion, respectively).

system's center of mass at perihelion and aphelion; and ε is the distance between the center of the Sun and the solar system's center of mass.

Diagrams for calculating the two locations for the planets are shown in Figure 5.6.

Using the above equations, we calculated the distances for CP and CA, and from them, changes in the solar constant, whose value is inversely proportional to the square of the distance to the Sun. It was conventionally assumed that at the initial moment ($t = 0$), the apses line coincides with the axis that aligns the Sun, Jupiter, and Saturn (when both planets are on one side of the Sun). The ellipticity of the orbits and non-coincidence of their planes was not taken into account. Figure 5.7 shows the results of the calculations in which cyclic variations of the solar constant with a period of 60 years can be clearly seen. The range of these variations is approximately 2.4% (about 33.6 W/m^2).

Pogosyan and Turketti (1970) estimate that only 17% of extra-atmospheric solar radiation is absorbed by the Earth's surface and the atmosphere, due to spreading over the spherical area of the Earth and reflection by the atmosphere. Hence, the range of anomalies in the amount of heat absorbed by the atmosphere and the Earth's surface due to dissymmetry of the solar system in a 60-year rhythm is about 6 W/m^2.

Budyko (1969) concluded that a change in solar radiation of several tenths of a percent is sufficient to trigger a significant change in Earth's climate. Note that the impact on the climate system of radiation reduced by greenhouse gas concentration in

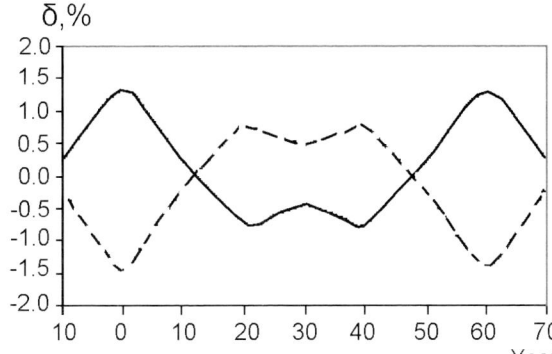

Figure 5.7. Variations in the intensity of extra-atmospheric radiation in January (solid line) and in July (dashed line) during a 60-year cycle (% of the corresponding average value).

2000 (compared to 1750), averaged over the Earth, is only $2.5\,W/m^2$ (Izrael *et al.*, 2001), which is 2.3 times less than the value given above. Kondratyev (2004) shows that an increase in the radiation balance of Earth's atmosphere determined by the supposed doubling of CO_2 concentration comprises only about $4\,W/m^2$, which is also less than the variations in absorbed solar radiation during a 60-year cycle that we have calculated.

Figure 5.7 shows that Earth's elliptical orbit causes opposite signs in solar constant anomalies at perihelion and aphelion, while its values on average for a year barely vary. Because the effect of insolation anomalies on the ice cover and on Earth's climate (especially accounting for polar day and night conditions) depends on the time of the year, interseasonal differences in the signs of anomalies will be accompanied by changes in the climate system during a 60-year cycle.

It is important to note that the role of a 60-year cycle is maximal in regions where the amplitude of seasonal insolation variations and their effect on the temperatures of the underlying surface and the adjacent air layers are most pronounced. These conditions are typical of high and marginally temperate latitudes because of the ice and snow cover. In low latitudes where the state of the underlying surface changes little during the year, the impact of this cycle is minimal. Analysis of air temperature observations in different climatic zones confirms this result (see Table 4.1).

The patterns discussed result in the occurrence of additional differences between solar heat influxes to the Southern and Northern Hemispheres and corresponding anomalies in air temperature gradients between the hemispheres that bring additional anomalies of heat, moisture, and mass exchange in the atmosphere and ocean. The values and the signs of these anomalies exhibit a 60-year rhythm. This suggests that a 60-year cycle of solar radiation anomalies may result in 60-year fluctuations of climatic characteristics and a corresponding cycle of sea ice extent fluctuations in the Arctic Seas.

It should be noted that the results presented above characterizing fluctuations in the solar constant during a 60-year cycle cannot be used as a tool for specific calculations. The fluctuations are not referenced to a specific year, and the specific orientation of Earth's orbit relative to the locations of Jupiter and Saturn, as well as the eccentricities of their orbits, were not taken into account. In addition, the distances to Jupiter and Saturn and the periods of their revolutions were not sufficiently accurate. Also, there may be effects from the outer planets of the solar system (Uranus, Neptune). The main objective here is to show that there is a sound logical basis for believing that "50–60 year" cycles of climate fluctuations and ice extent in the Eurasian Arctic Seas may be due to astronomical factors. Validating the estimates of the influence of solar-constant changes on climate will require considerable further research, including physical-mathematical modeling to determine the spatial-temporal characteristics of the distribution of solar insolation anomalies over the entire Earth.

6

Assessment of possible changes in air temperature and sea-ice extent in the Arctic Seas in the twenty-first century

6.1 BRIEF REVIEW OF THE METHODOLOGIES APPLIED

Attempts at super-long forecasts of ice extent for some regions of the Arctic Ocean have a long history. They have mainly focused on the sub-Atlantic part of the Arctic, because long series of ice observations are available for that area. Maksimov (1955) made the first forecast, based on space-geophysical factors and, primarily, a century-long cycle of solar activity. According to this forecast, maximum ice extent in the region was expected in 1990. This forecast, however, was not correct (nor were other forecasts based on the "100-year" cycle of solar activity). In updating Maksimov's forecast, Latukhov and Sleptsov-Shevlevich (1995) based their work on the relationship of ice extent and the magnetic perturbation index K_p, which also reflects the 100-year cycle; they predicted that maximum ice extent in 2000–2020 would be comparable to the conditions of the early twentieth century. As we can see now, this prediction was also incorrect. The results of reconstructing ice extent changes in the eighteenth and nineteenth centuries presented by these authors suggests some doubt about the validity of their methodology. According to their results, the second half of the nineteenth century was distinguished by decreased ice extent in the sub-Atlantic region of the Arctic. However, data collected by Norwegian scientists (Vinje, 2000) showed that during this period, increased ice extent was observed there.

Rudyaev et al. (1985) provided a more realistic climate forecast for the first half of the twenty-first century based on 65-year and 33-year cycles in the Earth's rotation speed. This forecast predicts maximum warming between 2005 and 2010, followed by a period of cooling that will last until the middle of the century. It seems likely that the Earth's rotation speed is an important indicator of climate change as large-scale anomalies of air temperature, atmospheric circulation, and ice extent are statistically connected with it. The angular speed increases during periods of climate warming and decreases during periods of cooling. The proposed physical mechanism for this relationship is based on the assumption that large-scale anomalies in atmospheric circulation lead to changes in west-to-east atmospheric circulation in temperate latitudes, with corresponding changes in the integral moment of wind tangential stress at Earth's surface and resulting significant changes in the planet's angular rotation speed after about 10 years. The correlation coefficient between the mean annual values of Earth's angular rotation speed in the twentieth century and the NAO index showing west-to-east air transport about 10 years later is 0.85 (Gudkovich et al., 2004).

Co-authors of the monograph (Climatic Regime, 1991) detected a 15-year phase shift between fluctuations in ice extent and pole deflection, and predicted ice extent in the Arctic Ocean and the Kara Sea for 1990–2005. Ice extent was expected to reach its maximum by 2005 and not to exceed the anomalies observed in the twentieth century. However, by 2005, no significant increase in average ice extent was observed. The same study also touches upon the possible influence of dissymmetry of the solar system's center of mass on average air temperature, and suggests some increase in temperature by 2005.

Gudkovich and Kovalev (2002b) forecast average anomalies in total ice extent of North Asian shelf seas by 5-year periods through the middle of the twenty-first century. The forecast is based on a physical-statistical model that incorporates the long-period cyclic changes and the linear trend of the twentieth century. It assumes a gradual growth in total ice extent during the first half of the twenty-first century, with a maximum expected in the middle of the second quarter (2025–2050) of the century. These changes in ice extent are within the framework of actual fluctuations observed in the twentieth century.

Forecasts of twenty-first century changes in Arctic Ocean ice extent developed by the supporters of a decisive role for greenhouse gas accumulation in climate change and based on coupled ocean-atmosphere models, are discussed in Section 5.1.

6.2   ASSESSMENT OF EXPECTED CHANGES IN AIR TEMPERATURE AND SEA-ICE EXTENT BASED ON CYCLIC FLUCTUATIONS

The pronounced character of the 60-year cycle in air temperature variations in the Arctic (Figures 4.1 and 4.2) provides a basis for a long-term forecast of climate change in the Arctic for the coming decades. The period of dominant positive air temperature anomalies began with the first Arctic warming from 1922 to 1954, followed by a cold period from 1955 to 1980. The last stable warming period began in the mid 1980s and continues today with a maximum displayed in the end of twentieth—beginning of twenty-first centuries. The amplitude of 0.6°C and the phase from Figure 6.1 can be used to forecast the future. Based on these estimates, it can be expected that after the first decade of the twenty-first century, the Arctic background temperature will start to decrease and reach a minimum by 2030-2035, after which we should expect a transition to the next warming event (Figure 6.1).

A forecast for possible twenty-first century changes in Arctic ice extent based on natural cyclic changes is shown in Figure 6.2. This forecast takes into account the main components of our derived long-term variability in twentieth-century ice extent:

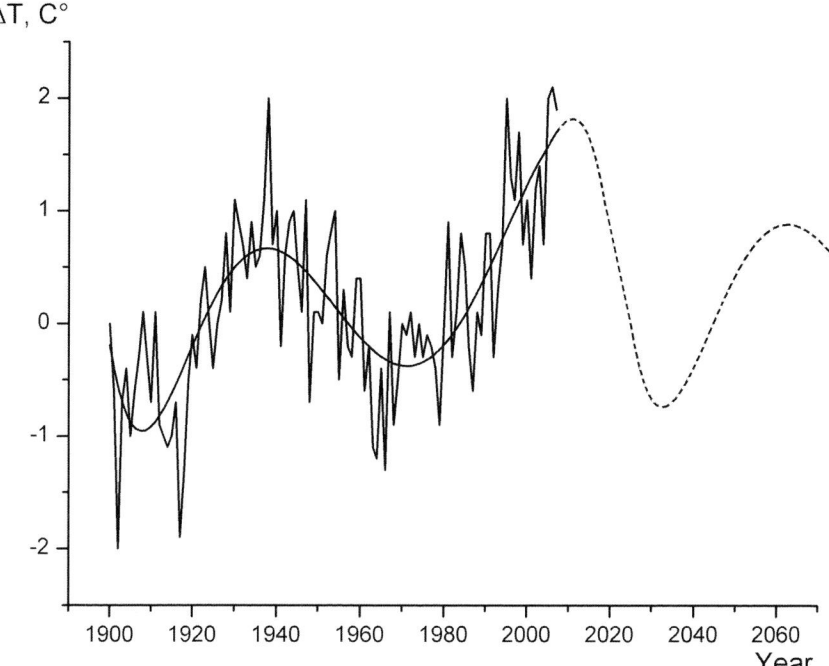

Figure 6.1. Changes in the anomaly of mean annual air temperature in the 70°N–85°N zone during 1900–2007 (solid line), and its background forecast (dotted line).

a "60-year" cycle, a linear trend in the second half of the twentieth century, and a "20-year cycle" for the seas of the western region. For updating the average duration, amplitude, and initial phase of the "60-year" cycle, harmonic processing of the polynomial trend ordinates of total ice extent was implemented separately for the western and eastern seas. Amplitudes used for the western and eastern seas were $210 \cdot 10^3$ km^2 and $70 \cdot 10^3$ km^2 respectively.

As shown in Figure 6.2, it is predicted that in the twenty-first century, oscillatory (rather than unidirectional) ice extent changes in the Arctic Seas will continue. During the 2020s–2040s, an increase in ice extent is projected, with a maximum around 2030 in the eastern seas and around 2035 in the western seas. The next maximum falls at approximately 2090–2095 (Karklin *et al.*, 2001; Gudkovich *et al.*, 2002b, 2005).

An important factor for shipping conditions in the Arctic Seas is the duration of the period that allows unescorted navigation. Table 6.1 shows average durations of through voyages without icebreaking escort along the NSR (from the Kara Sea to the Chukchi Sea), with the average varying from 0 to 35 days.

During the periods of increased ice extent (1962–1983), icebreaker escorts were needed for 50% of cruises along the NSR in order to provide through voyages, while in periods of decreased ice extent (1933–1961 and 1984–2004), icebreaker escorts were required in only 17% and 14% of voyages, respectively.

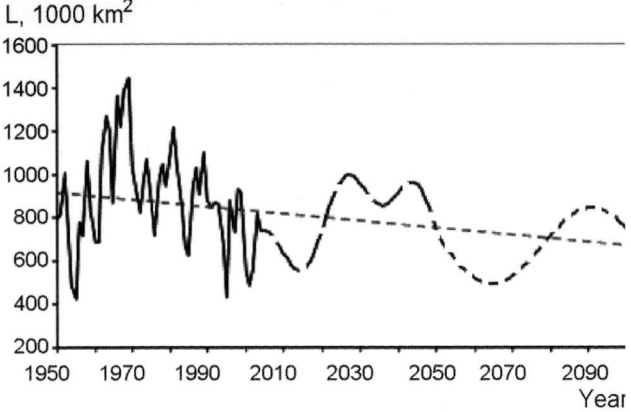

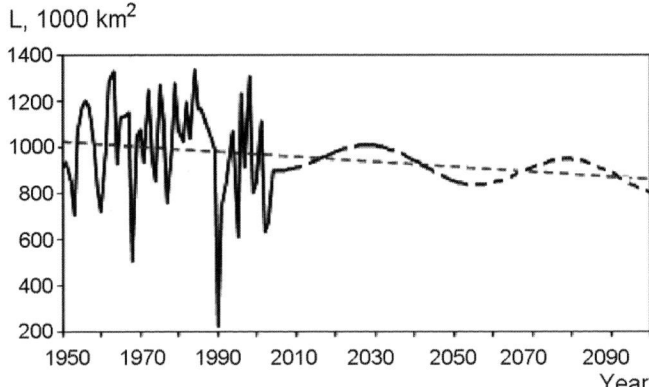

Figure 6.2. Forecast of climatic changes for the total area of ice extent in the western (a) and eastern (b) Eurasian Arctic Seas for the twenty-first century, taking into account the linear trend in the second half of the twentieth century.

Table 6.1. Average duration of unescorted through navigation along the NSR (depending on the total ice extent of the Arctic Seas).

Total ice extent gradations (%)	Average duration of through voyage without icebreaker support (days)
≥12	0
+2 to +11	9
−4 to +1	13
−9 to −5	26
≤ −10	35

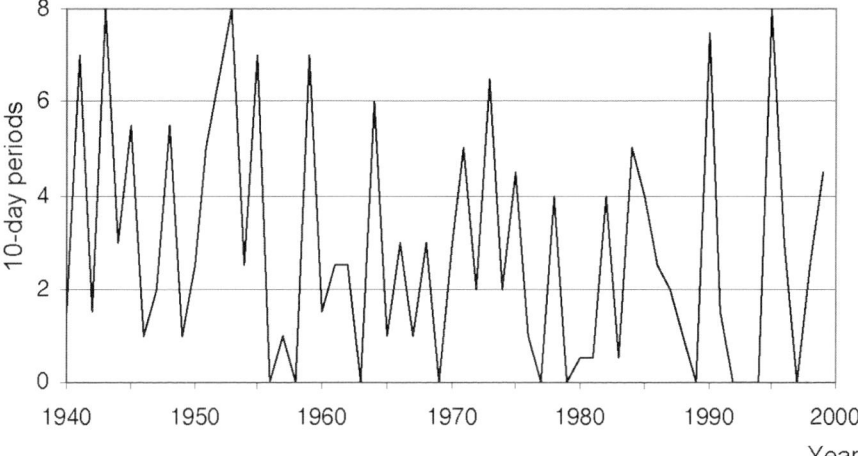

Figure 6.3. Number of periods of unescorted through navigation for the Russian ice-strengthened (UL′) ice class ships along the NSR (10-day periods)

This is confirmed by actual data on the duration of periods of unescorted through voyages along the NSR for 1940–2000, as shown in Figure 6.3; even during the last period of Arctic warming, there were years when unescorted through navigation was impossible. Thus, ice conditions expected at least during the first half of the twenty-first century suggest a continuing need for icebreaker support of marine operations in the Arctic.

6.3 SEA-ICE VARIABILITY DURING 2003–2008

The Russian edition of this monograph published in 2007 presents analyses of sea ice data dating to 2003. For the English edition of this monograph, we have extended our studies to analyze significant and sometimes extreme (2007) changes observed in the Arctic since 2003, and we decided to extend our studies and explain these changes based on previously discussed theories and hypotheses. Some of the published works were not reflected in the Russian edition of the monograph and the authors have tried, as far as possible, to fill this gap. In this section we aim to investigate the cyclic variability of multiyear changes in the various Arctic climate system hydrometeorological parameters, in order to assess their reliability as well as their role in the total dispersion of ice extent. Cyclic variability data on scales from decades to centuries (in the Holocene) and tens of millennia (late Pleistocene) are now available for a number of Earth's climate indexes (Monin and Sonechkin, 2005).

As discussed in the second chapter of this monograph, changes in the ice extent of the Eurasian Arctic seas in the twentieth century exhibited a long-term negative trend accompanied by cyclic variations with periods of 50–60, about 20, 9–12, 7–8, and fewer years. However, the variability of sea ice extent in the western region

(Greenland to Kara Seas) was mainly influenced by a long-term trend and low-frequency cycles (about 60 and 20 years), whereas in seas of the eastern region (Laptev, East Siberian, and Chukchi), the influence of higher-frequency cycles prevailed (10 years and less).

Studies by many scientists confirm the above cycles in interdecadal variability of ice cover area and other characteristics of Earth's climate system. These studies include the following.

Alekseev et al. (2003) reveal differences in the occurrence of warming and cooling epochs throughout the twentieth century as well as their association with atmospheric circulation. Monin and Sonechkin (2005) analyze large-scale climatic cycles related to alternation of glacial and interglacial epochs in Earth's history with special attention to 60- and 180-year cycles. The 180-year cycle, which we conventionally refer to as 200-year cycle, may be responsible for intra-century trends in climate change.

Klyashtorin and Lyubushin (2006) obtained 60-year oscillations in global surface air temperature based on instrumental measurements for the last 140 years. Using mean annual air temperature reconstructed from the oxygen isotope O^{18} concentration in ice cores from the Greenland glaciers for the last 1500 years, these authors carried out a spectral analysis that supported the existence of 60-year and 200-year cycles in climate changes. The results of these studies suggest that natural cycles, rather than greenhouse gases, may be the dominant factor in variability of the Earth's climate.

Proshutinsky and Johnson (1997) and Polyakov and Johnson (2000) present studies of changes in the Arctic Ocean regime associated with the Arctic Oscillation (AO), including cycles lasting 10–20 years. Minobe (1997) provides the results of studies of air and water temperature changes for different regions of the Pacific Ocean, North America, and the part of the Indian Ocean that adjoins the Asian continent from the south. This author concludes that in all the above regions, a characteristic feature of climate change is cyclic oscillations that last for 50–70 years and that are associated with alternating warm and cold epochs. The oscillation phases in most of these regions virtually coincide with similar Arctic cycles. The periods of 1870–1889, 1925–1947, and 1977–1990 qualify as warm epochs in these regions, and 1890–1924 and 1948–1976 as cold epochs. The oscillation phase is opposite only in the western part of the Pacific Ocean (Japan), which is due to the influence of the rear part of the Aleutian Low, where atmospheric pressure varies in accordance with alternating warm and cold epochs. The Aleutian Low deepens in the warm epochs and is partly filled during the cold ones. As will be shown below, climate changes in the northern part of the Atlantic Ocean have similar features.

Our follow-up studies were aimed at revealing spatial-temporal features of approximately 60-year cycles in the Northern Hemisphere during the first and second warming epochs in the twentieth century. The differences in these epochs may be associated with a longer cycle that is evident in the linear trends of variations in different climate-system indicators. Figure 6.4a, b (see color section) shows the distribution of surface air temperature differences averaged for the winter and summer periods of 1980–2000 compared to 1930–1950 (Frolov et al., 2009). From the first to the second warming epoch, the temperature increased over Greenland, the mid-latitudes of Eurasia and North America, the near-Pacific Arctic, and the northern Pacific Ocean in winter, and over the Arctic basin, the Siberian shelf seas, the northern Pacific Ocean, Central Asia, and southern Siberia in summer. However, in the same period, winter air temperature dropped significantly in the North Atlantic and over a major part of the Arctic Ocean, including areas adjoining the Baffin Sea and the Arctic seas of the Eurasian shelf from the Barents Sea to the East Siberian Sea. As shown in Section 2.2, a decrease in air temperature was accompanied by growing ice extent in the Barents Sea in winter, which amounted (on average for October–February) to about 130,000 km^2 in 50 years. Decreased air temperature in this period also occurred in the summer half of the year in the region west of Greenland, over the North Atlantic, Western Europe, eastern Siberia and the southern Asian continent.

Considering that cooling was recorded at the same time in the Antarctic (Gudkovich *et al.*, 2008), it should be acknowledged that although "global warming" occurred as a global average, this was not uniform spatially or temporally, and cooling was recorded over large areas of our planet in both winter and summer during the last decades of the twentieth century. Note that the value of anomalies characterizing an epoch depends on the period for which the climatic "norm" is determined.

It was noted in Sections 4.1 and 4.2 that air temperature at mid and high latitudes primarily depends on dynamic processes in the atmosphere (Alekseev, 2000; Alekseev *et al.*, 2003; Vorobiev and Smirnov, 2003). They influence air temperature due to both advective processes and the impact of cloudiness, which depend on the type of baric system in play. In winter, this influence is particularly high in areas where anticyclones are common. Weakening of anticyclones results in increasing temperature and cloudiness. Variation in cloudiness is one of the main causes of climate change is indicated by Sherstyukov (2008).

Maps showing differences in mean sea level pressure for the winter and summer periods (not given here) for 1980–2000 compared to 1930–1950 and characterizing changes occurring from the first to the second epoch of warming confirm the pattern mentioned above (Frolov *et al.*, 2009). In the winter half of the year, atmospheric pressure over the Arctic during the last warming epoch was much lower than it was during the first warming. Pressure also dropped in the Siberian, Canadian, Greenland, and Arctic Highs. Due to the deepening and southward displacement of the Icelandic Depression (the most important atmospheric center of action) and a significant pressure increase over the North Atlantic, intense zonal transport in the atmosphere shifted from the high- to the mid-latitudes. This was a direct cause of a significant mid-latitude air temperature increase over the Eurasian and North American continents, where seasonal anticyclones are commonly located at this time of the year (Klimenko, 2007; Wallace *et al.*, 1995). Temperatures dropped where thermal conditions were influenced by the rear areas of baric depressions (Baffin Sea, North Atlantic, Barents Sea) (Alekseev, 2000, 2004; Klimenko, 2007; Wallace *et al.*, 1995).

A similar map characterizing the summer half of the year points to a major atmospheric pressure drop over the Arctic basin in the last warming epoch along with a slight increase over Eurasia and northern regions of the Atlantic and Pacific oceans. This created a favorable temperature background (for ice decrease) over the Eurasian Arctic seas and an unfavorable one over the northwest Atlantic Ocean, including the Baffin Sea.

Walsh *et al.* (1995) describe a major reduction in atmospheric pressure at sea level over the Arctic near the end of the 20th century. Thus, air temperature variations during the time period between the warming epochs can be accounted for by corresponding variations in the average pressure fields that characterize atmospheric dynamics. These changes correspond to climatic variations in the *condition of polar (circumpolar) vortices* (Dmitriev and Belyazo, 2006; Gudkovich *et al.*, 2008). It is known that cyclonic rotation of the troposphere and the lower atmosphere from west to east around the poles is associated with polar vortices. In the lower layers of the atmosphere, over the Arctic Basin in winter, cyclonic vorticity changes its sign: an Arctic High forms here. In summer, a cyclonic field is commonly found near the surface (Dolgin, 1968; Anon. (E)). A somewhat similar pattern of general atmospheric circulation is observed in the Antarctic, through its different distribution of land and sea, as well as the presence of a thick Antarctic glacier, result in certain differences (Anon. (D)).

The intensity of circumpolar vortices varies within a year, driven by seasonal air temperature gradient variations between low and high latitudes; in the winter period in both hemispheres, atmospheric circulation intensifies, and in the 6-month summer period it weakens. The association between the state of polar vortices and air temperature is quite different in the climatic variability of warming and cooling epochs. In warming epochs, atmospheric pressure and geopotential values within the troposphere and the lower stratosphere decrease in the zone of polar vortices. This results in intensification of zonal flows in the atmosphere of mid-latitudes, which are apparent in indices of general atmospheric circulation, such as the North Atlantic Oscillation, the Arctic Oscillation, high-latitude zonation, and others in the Northern Hemisphere, and the South Polar Oscillation in the Antarctic. In cooling epochs, zonal flows become weaker (Gudkovich *et al.*, 2008). It should be noted that the Arctic High weakens with an intensifying northern polar vortex; with weakening of the vortex the Arctic High strengthens (Dmitriev and Belyazo, 2006).

The patterns of variation in the intensity of zonal flows in the atmosphere of mid-latitudes described above are confirmed in Section 4.2. Particularly, variation in the mean annual zonality index is shown to express the difference in atmospheric pressure at sea level between 40°N and 65°N during the 20th century (Figure 4.9). In addition to a characteristic increase in the index from cold to warm epochs, the pattern reveals an intensification of zonal transport from the first to the second warming epoch due to the fact that the belt of intensified zonal transport in the atmosphere displaces from high to mid-latitudes as a result of the extension and the deepening of the polar vortex.

What causes these variations in atmospheric circulation?

Variations in circumpolar vortices may be caused by both external and internal factors. Among the internal factors, until recently, most climatologists placed major emphasis on the effect of accumulating anthropogenically generated greenhouse gases (mainly CO_2) in the Earth's atmosphere. Section 5.1 provides reasons why the "greenhouse theory" has weak foundations.

A series of papers by G. V. Alekseev and his co-workers examines the low-frequency cyclic oscillations of climate with a period of 60–80 years. In these papers, it is presumed that the first twentieth-century warming period was character-ized by higher surface air temperatures in the near-Atlantic Arctic, and the second warming period exhibited higher surface air temperatures in the near-Pacific region and other latitudinal zones (Alekseev, 2003; Alekseev *et al.*, 2003; Alekseev and Ivanov, 2003). Because anthropogenic emission of greenhouse gases to the atmo-sphere in the first warming epoch was far less, and became apparent only by the time of the second warming, the authors made a presumption: warming in the first half of twentieth century was caused by natural oscillations of the climate system, and the last warming "cannot be accounted for without regard for the anthropogenic fac-tors." Greenhouse gases resulting from burning fuel and, partly, from emissions by volcanic activity (Katsov, 2003; Johannessen *et al.*, 2004; Vinnikov *et al.*, 1999) were recognized as such factors by those who carry out coupled models of the atmosphere and the ocean. Note that, based on temperature diagrams provided in IPCC reports (2001, 2007), the strongest recent volcanic eruptions only impacted the Earth's climate for a maximum of 3 years, and thus they cannot be the cause of climate changes on the scale of decades.

The Report of the Nongovernmental International Panel on Climate Change (NIPCC) (Singer, 2008) criticized the main IPCC conclusions regarding the intensi-fication of anthropogenic global warming in recent years. The NIPCC report argues that the magnitude of global warming is essentially overestimated due to the influence of urban heat islands on measured surface air temperature. Rapp (2008) discusses the inadequacies of the surface temperature measurement system in some detail. Never-theless, it cannot be argued that the Earth has not warmed significantly in the 20th century. A major problem for climate models is how to deal with putative increases in humidity resulting from increases in global temperature due solely to increased CO_2. Most models treat humidity as a global average, and since water vapor is a powerful greenhouse gas, this greatly amplifies the temperature increase due to increased CO_2. However, Lindzen (1997) emphasized that the degree of water vapor feedback as a heating force in any region depends on the absolute humidity. In desert regions with very low absolute humidity, an increase in humidity provides a significant heating force. However, in regions with high absolute humidity, an increase in humidity provides a very modest heating force. Tropical regions that already have high humidity, do not gain much additional heating from an increase in humidity.

Climate models assume that the main factor affecting the atmosphere is the greenhouse effect of carbon dioxide, but they do not account for the Earth's inter-decadal climate changes that affect the evolution of circumpolar vortexes discussed above. Moreover, Gudkovich *et al.* (2008) averaged the calculations for fields of atmospheric pressure from five models (HAD, CNRN, EHAM, GFDL, INM) and found that they significantly overestimate atmospheric pressure over the Arctic basin during climate warming periods. This contradicts the finding by Vize (1944b) and confirmed in subsequent years that air temperature and ice extent in the Arctic seas are fundamentally dependent on the degree of development of the Arctic High. The model calculations also contradict the observation that, as shown above, air temperature anomalies are primarily dependent on dynamic processes in the atmosphere. Errors in the calculated temperature would undoubtedly lead to major inaccuracies in model predictions of ice cover conditions and other climate characteristics.

In our opinion, a reliable argument contesting the decisive role of the anthropogenic factor in climate change is the decrease in winter air temperature over large regions of the Arctic. It is known that, in the winter season, long-wave radiation plays a decisive role in the heat balance of polar seas. Long-wave outgoing radiation could theoretically be affected by the concentration of greenhouse gases in the atmosphere, which strongly increased by the end of the twentieth century (IPCC, 2007). Nevertheless, there did not seem to be any effect on the atmosphere at high latitudes; instead of warming, cooling was recorded over vast spaces. Even accounting for the positive temperature anomalies recorded in the first decade of the twenty-first century, average air temperature in the second warming epoch was not higher than in the first one.

Water vapor played a role in the air temperature increase at the end of the twentieth century in regions where seasonal anticyclones occur in winter; note that water vapor has a greater influence on effective radiation of the atmosphere than greenhouse gases of anthropogenic origin. An increase in cloudiness and water vapor content in the atmosphere over continents (but not over desert continental regions) in this period was due to a decrease in atmospheric pressure, which caused more intense cyclonic activity. This is confirmed by a corresponding growth in river runoff (see Section 4.7). These factors have a major influence on global air temperature as does the larger area occupied by the mid-latitudes compared to the high-latitudes. As a result, unlike the pattern for air temperature in the Arctic, the first warming epoch is less prominent in global temperature records than the second one.

Short-wave solar radiation is the most significant summer-season forcing, or, more precisely, the part of it that depends on albedo and absorption by the ice cover and the sea. Due to changes in albedo not related to greenhouse gases of anthropogenic origin, this heat balance constituent can vary by several dozen W/m^2 in polar regions, or one order of magnitude greater than the most optimistic assessments of the influence of greenhouse gases.

As an alternative to the "greenhouse theory" as a main cause of climate change in the late twentieth and the beginning of the twenty-first centuries, the effect of solar activity (SA) on atmospheric processes attracts considerable attention. Section 5.2 provides a review of papers (mainly by Russian scientists) on the relationship in time between changes in ice extent and other characteristics of the climate system with SA parameters (mainly Wolf numbers). Luk'yanova (2007) offers interesting facts related to these issues that have largely been discovered by non-Russian scientists.

Satellite measurements have brought new understanding of the influence of SA on the total solar irradiance (TSI) to the Earth (Frolich and Lean, 1998) and its climate (Douglass and Clader, 2002). Gudkovich *et al.* (2005) suggest a positive linear trend in SA (Wolf numbers) in the twentieth century as a possible cause of corresponding climate changes in the Arctic. In Figure 6.5, borrowed from Soon (2005), changes in annual air temperature anomalies north of 62°N (Polyakov *et al.*, 2003) are compared with TSI values estimated by Hoyt and Schatten (1993), as well as with CO_2 content in the atmosphere from 1875–2000. The variation of temperature matches the TSI curve far better than it matches the CO_2 curve. However, the Hoyt and Schatten model for TSI is just one of many, and other models lead to very different patterns for TSI vs. year. Furthermore, climate modelers would argue that the temperature curve in the second warming epoch represents the continuation of the first warming epoch, interrupted by a period from about 1940 to about 1980 when increasing aerosol concentrations outweighed the effect of increasing greenhouse gases. Therefore, Figure 6.5 is just one representation of many that could be derived. Nevertheless, if Figure 6.5 were taken at face value, the temperature and TSI varia-

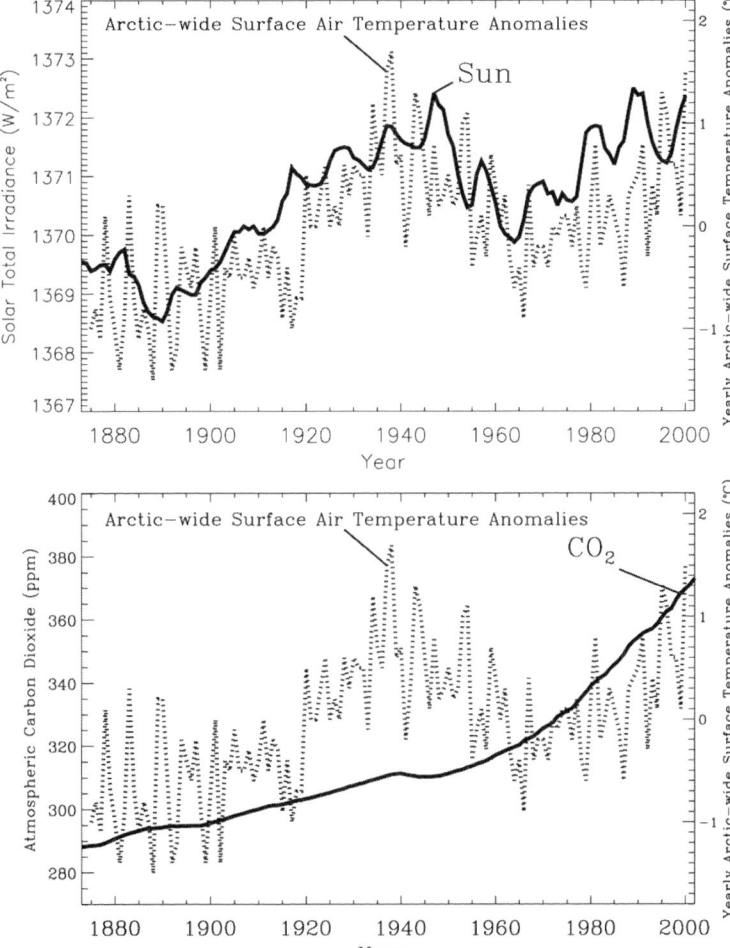

Figure 6.5.
Annual-mean
Arctic-wide air
temperature
anomaly time
series (dotted
line) correlated
with estimated
total solar
irradiance (solid
line in the top
panel) from the
model by Hoyt
and Schatten,
and with the
mixing ratio of
atmospheric
carbon dioxide
(solid line in the
bottom panel)
(Soon, 2005).

tion charts would suggest the presence of both a positive "100-year" trend and quasi 60-year cyclic oscillations. This is also corroborated by the correlation coefficients between annual average air temperature in the Arctic, TSI, and SA anomalies given in Hoyt and Schatten (1993). With 10-year smoothing, the coefficients are 0.89 (with TSI) and 0.47 (with CO_2). Hence, according to this particular model, the main cause of climate changes in the Arctic would be the dependence on TSI rather than buildup of greenhouse gases. However, there is no way to test the Hoyt and Schatten model, and other models for TSI exist with quite different results. While Figure 6.5 is suggestive, the fact remains that we really do not know how TSI varied prior to the advent of satellite measurements around 1980. Figure 6.5 demonstrates that the form of the variability of Arctic surface temperatures during the 20th century resembles the variability of the Hoyt and Schatten model for TSI. This is suggestive that variations in TSI may have been an important factor in 20th century climate change. Though the total variance of TSI from 1880 to 2000 according to Hoyt and Schatten was $3-4\,W/m^2$, the simple spreading of this flow over the spherical area of the Earth is incorrect. As we show in this work, a significant part of TSI variance influences the high-latitude regions. Furthermore, as was noted in Section 5.4, Budyko (1969) concluded by calculations that solar constant variations of several tenths of % are sufficient to induce essential climate changes.

In seeking a relationship between solar variability and climate change, we may consider TSI and SA. The connection between TSI and climate is direct; TSI represents the fundamental heat input from the Sun that drives our climate. However, although SA represents fundamental aspects of the dynamics of the Sun, its connection to the total power emitted by the Sun is not quite clear. SA includes energetic particle emission, electromagnetic emission in the UV and higher frequency ranges and magnetic fields. It is manifested in the Earth's phenomena in the form of polar lights, magnetic storms, radio-communication blackouts, etc. A number of different indices are used to measure the level of SA, particularly sunspot indices (Wolf number, etc.), the intensity of solar wind, and various magnetic indices. Even though variations in TSI associated with changes in SA may be small, the impact on higher latitudes is significantly amplified by the interaction of charged solar wind particles with the Earth's magnetic field. As shown in our work, evidence exists that variability of SA is connected to Arctic climate variations. In addition, we have also shown in Section 5.4 that the interaction of the gravity fields of the Sun and the solar system planets ("dissymmetry of the solar system planets"), by which we mean a displacement of the Sun's center relative to the center of the system mass, can produce seasonal changes in the solar input to Earth, which would affect the climate of higher latitudes on a 60-year cycle.

The cause of a "100-year" trend in SA that may be associated with a 200-year cycle (Westbrook, 1998; Bashkirtsev and Mashnich, 2004; Raspopov, 2004) has not yet been reliably determined. However, the 60-year SA cycle (Fritz cycle) is probably due to the influence of "dissymmetry of the solar system" (see Section 5.4), which changes the distance between the Earth and the Sun on a 60-year cycle (Gudkovich *et al.*, 2005). Over the longer term, if we can confirm existence of a 60-year cycle in TSI, this might confirm the theory, and would also provide a basis for explaining the opposite signs found in the 60-year climate cycles of the Arctic and the Antarctic.

Monin and Sonechkin (2005) provide some support for such an explanation of climate variations. They consider the cause of 60-year climate variability to be a triple cycle of solar magnetic activity (the Hale cycle) that is shown to last about 60 years by wavelet analysis of a number of hydrometeorological parameters. "In such a triple loop, the Sun follows a trajectory around the center of inertia in the form of a slightly unlatched trefoil, always behind the center of inertia at a distance of slightly more than the diameter of the Sun" (Monin and Sonechkin, 2005, p.16). The same complex quasi-periodic motion also includes a cycle that averages 179 years in length. It is related to "... variations of the solar radiation incoming to the Earth, which change in many respects due to the gravitational interactions between the Sun and planets, especially Jupiter and Saturn (Monin and Sonechkin, 2005, p. 43).

Isotope analyses of ice cores drilled from glaciers in the Antarctic and Greenland using radionuclides of cosmic origin (^{14}C and ^{10}Be) allowed reconstruction of Wolf numbers as far back as the middle of the ninth century AD (Usoskin et al., 2003; Solanki et al., 2004). Figure 6.6a (see color section) presents the results of several versions of this reconstruction. Although there is considerable variance from model to model, the models all suggest that sunspot numbers appear to have been lower on average for a thousand years prior to the 20th century. While we do not have reliable models to connect sunspot activity to TSI and climate change, the fact that sunspot activity appears to have increased significantly in the 20th century suggests that climate change in the 20th century may, at least in part, be related to solar changes.

There have been many attempts to estimate the historical surface temperatures over the past few hundred years or in a few cases, as far back as two millennia. Studies based on temperature proxies (tree rings, ice cores, coral terraces, pollen counts, etc.) have estimated temperatures in various regions for various time periods. Mann et al. (1998, 1999, 2003, 2004, 2008) attempted to integrate the entire array of proxies into a single cohesive reconstructed estimate of the global average temperature over the past two millennia. While the results of such reconstructions by Mann and others vary considerably from study to study (see Figure 6.6b, see color section), they all lead to a common morphology of a temperature profile for two thousand years, followed by a sudden sharp rise in the 20th century. There is moderate evidence of the so-called "Medieval Warm Period" or a "Little Ice Age" in these results. It should be noted that the red curve at the far right of Figure 6.6b (see color section) (CRU instrumental record) is grossly exaggerated in vertical height. The figures indicate a global temperature rise of 1.3°C in the past century—about double the accepted value. This result has served the needs of global warming alarmists who view the sudden temperature rise in the 20th century after two millennia of small variations, as evidence of the anthropogenic impact on climate change. McIntyre and McKitrick (2003, 2005, 2006, 2007) and Wegman, Scott and Said (2006) found errors in the data reduction procedures used by Mann et al., and dubbed the temperature profile obtained by Mann et al. in derisive terms as the "hockey stick." Rapp (2008) describes this controversy in considerable detail. While Figure 6.6b (see color section) appears to underestimate the climate variations associated with the so-called "Medieval Warm

Period" or a "Little Ice Age" and it exaggerates the rise in the late 20th century, nevertheless, the comparison of Figures 6.6a and 6.6b (see color section) is suggestive that there may be a solar connection to long-term climate change over the past millennium.

The data in Appendix A show that the cover area of sea ice decreased during the period from 2000 to 2008. A small positive ice extent anomaly was recorded only in the eastern region during two of the nine years (2001 and 2004). The year 2007 appeared to be the warmest, when the maximum air temperature, the minimum ice extent of the Arctic seas, and other extremes of the observation series were recorded. The highest anomalies were reported from the East Siberian and Chukchi Seas (Frolov, 2007; NCDC, 2007), i.e., from the region where the role of short-term fluctuations is significant (see Section 2.3).

The reviews referenced above and a paper by Dmitriev (2007) consider the hydrometeorological features of 2007 in detail, and their findings include the following phenomena. Zonal circulation in the atmosphere of the Arctic was abnormally high in 2007. Negative anomalies of surface atmospheric pressure in the Eurasian sector and positive anomalies in the American sector resulted in advection of warm air masses from the Pacific Ocean to the Arctic, which remained stable throughout most of the year. A positive anomaly of average annual surface air temperature in the Eurasian sector of the Arctic and in the zone north of 70°N reached 2.5–2.8°C (relative to the period 1961–1990).

These 2007 atmospheric processes created exceptionally favorable conditions for ice cover destruction in the Arctic seas studied herein, as well as in adjoining areas of the Arctic basin. Late onset of ice formation in the autumn of the preceding year, a low rate of ice cover formation, and intensified ice removal from the seas to the Arctic basin and further (to the Greenland Sea) in the winter of 2006/2007 resulted in the following: by the spring of 2007, the ice cover in these seas was mainly composed of first-year ice of decreased thickness but with inclusions of younger ice that formed in extensive flaw polynyas. Early onset of melting under these conditions resulted in rapid destruction of the ice cover, and stable drifting promoted ice removal beyond the boundaries of the seas. This was particularly evident in seas of the eastern region, which were in the zone of high baric gradients between the Arctic High and the Icelandic Depression extending far eastward.

As a result, as early as August 2007, high negative ice extent anomalies were recorded in these seas (Table 6.2). In September 2007, the ice edge in the East Siberian Sea sector approached 85°N, which had been never been recorded throughout the entire period of routine observations (i.e. since the 1930s). Intense melting and early disappearance of the ice cover from large areas of open water resulted in increased heating and freshening of surface water and late ice formation. All of this occurred

Table 6.2. Characteristics of ice extent anomalies in the Arctic seas in August 2007

Seas	Barents	Kara	Laptev	East Siberian	Chukchi	Beaufort
Anomalies (%)	−7	−3	−22	−76	−31	−35

against the background of the earlier positive anomaly of temperature (to $+1.5°C$) and greater thickness of the deep Atlantic water layer.

The major 2007 Arctic ice cover anomaly prompted many climatologists to revise their views on the intensity of Arctic ice area reduction connected with "global warming" due to greenhouse gases. In press releases to the mass media, a number of climatologists said that 2007 Arctic ice conditions pointed to an acceleration of the global warming process. In some of the interviews, it was predicted that the Arctic ice would disappear in the next five years (!).

Such views on climate change can be accounted for by the fact that some scientists, unfortunately, are apparently unaware of a very important principle regarding patterns of time variability in hydrometeorological parameters: the average absolute value of anomalies decreases with an increase in the averaging time. Consequently, extrapolation of the changes observed during short intervals to long periods is not appropriate.

The SA models based on cosmic radionuclides, such as that of Solanki *et al.* (2004) indicate a quasi-periodic behavior for SA indices. The Solanki model suggests that the probability of the persistent elevated solar activity for the next five decades is only about 8%. To the extent that the Arctic climate is driven by variations in SA, it would seem unlikely that warming observed in the 20th century will persist far into the 21st century. The fact that there have recently been short-term contractions of sea ice extent in seas of the eastern region of the Arctic cannot be extended to long-term trends.

As we pointed out previously, 2007 was an anomalously warm year for the Arctic, but one year does not create a trend. As it turns out, 2008 was colder than 2007 and ice extent in all seas in the eastern region of the Russian Arctic in August increased by a value exceeding $0.6 \cdot 10^6$ km^2 (Table 6.3), which corresponds to the reduction in ice cover area in all seas of the Eurasian shelf within the twentieth century (see Table 2.3). Significant increases in ice extent were also observed in the Arctic Basin and the whole Arctic Ocean. Assuming that by the end of September 2007 the area of the residual (first-year, second- and multiyear) ice in the Arctic Basin decreased to 2.92 million km^2 (according to weekly ice analysis provided by the Arctic and Antarctic Research Institute), since total ice extent at September 2008 was 3.47 million km^2, this represented an increase of 0.55 million km^2 (Frolov, 2008, 2009). The same estimates for the whole Arctic Ocean, available on a basis of the hemispherical ice analysis provided by the US National/Naval Ice Center (IICWG, 2008), were 3.98 million km^2 for the end of September 2007 and 4.66 million km^2 for the end of September 2008, an increase of 0.66 million km^2.[1]

---

[1] According to other estimates based on daily passive microwave SSM/I ice products (NSIDC Notes, 2007, 2008). the minimum ice extent for the Arctic Ocean of 4.67 million km^2 for 2008 was reached on 14 September 2008 and the minimum of 4.28 million km^2 for 2007 was reached on 16 September 2007 (an increase of 0.39 million km^2) . The difference between the ice charting analysis and the passive microwave estimates is mostly attributable to greater accuracy in ice analysis of the radar and visible satellite imagery used for the ice-charting purposes.

Table 6.3. Ice extent values recorded in the Eurasian Arctic Seas in August 2007 and 2008, 10^3 km^2

Sea	GS	BS	KS	LS	ESS	CS	Western seas	Eastern seas	Total
2007	304	42	236	144	0	0	582	144	726
2008	196	56	166	315	383	60	418	758	1176
Difference 2008–2007	*−108*	*+14*	*−70*	*+171*	*+383*	*+60*	*−164*	*+611*	*+450*

GS—Greenland Sea. BS—Barents Sea. KS—Kara Sea. LS—Laptev Sea. ESS—East Siberian Sea. CS—Chukchi Sea. The western seas encompass Greenland, Barents, and Kara, and the eastern seas Laptev, East Siberian, and Chukchi.

In light of the above it is interesting to consider the changes in the propagation of old (second- and multiyear) ice in the Arctic Basin in recent years that were not taken into account in Section 4.5. Figure 6.7 presents the changes of mean latitude of the old ice dominance boundaries (partial concentration 5 tenths and more) in late winter (March) for the three meridian sectors corresponding to the Laptev Sea, East Siberian, Chukchi and the adjacent areas of the Arctic Basin.

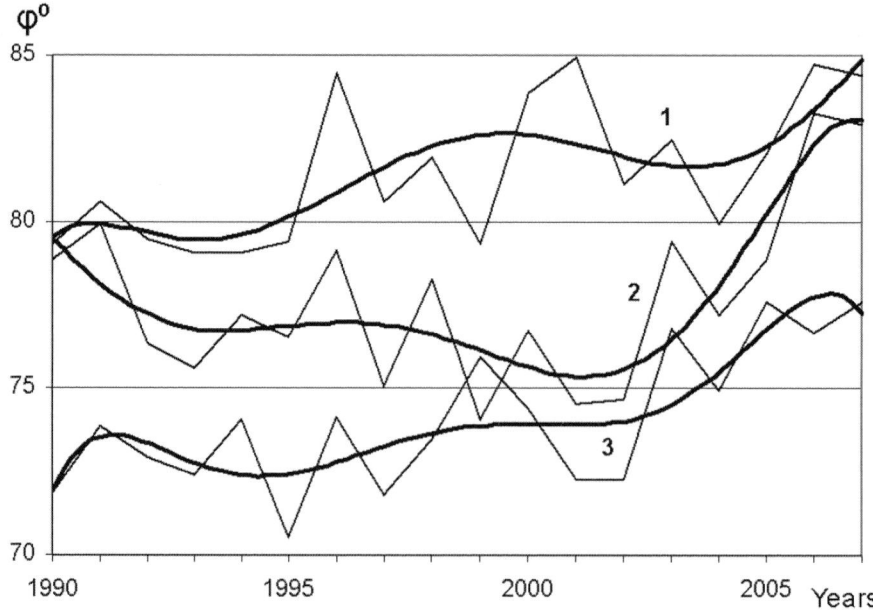

Figure 6.7. Mean latitude of the old ice dominance boundary (thin lines) and its approximation by a polynomial to the power of 6 (thick lines) in March 1990–2008 for the three meridian sectors corresponding to the Laptev Sea (1), East Siberian (2) and Chukchi (3) Seas.

This figure confirms that the gradual southward shift of the old ice boundary described in Section 4.5, which was observed in the second half of the twentieth century, continued at least until 2002. However, in the next 5–6 years, a substantial northward retreat of the old ice boundary occurred, which as noted above, was due to exceptionally favorable (for ice decrease) hydrometeorological conditions in the region associated with short-period cycles of atmospheric circulation. Calculations show that for period since 2002 the area of old ice in this sector of the Arctic Basin was reduced by approximately $2.106 \, \text{km}^2$. Such results do not contradict conclusions by Mahoney *et al.* (2008), although gaps in the data and 10-years running smoothing used in the analysis, distorts the temporal scale of fluctuations revealed by its authors. Similar fluctuations, although smaller in scale, were observed in the Laptev Sea and area of the Arctic Basin north of it in 1995–1997 (see Figure 3.3), as well as in the Beaufort Sea in 1998–2001.

Since 2007 there appears to have been a transition to a new phase of this oscillation, during which the boundary of the old ice started to move southward. This is confirmed by the changes that occurred from 2007 to 2008. In addition to the above changes in ice extent, the old ice boundary during the last year shifted southward which was revealed by the setting in September 2008 of a new North Pole drifting station (NP-36) onto a multiyear ice floe over 3 meters thick at $82°35'$N, $172°07'$E, where no ice was present in September 2007. In March 2009 the position of the MY ice boundary at the meridians of the East Siberian Sea was near $81°$N while during the previous year it retreated beyond the North Pole.

Arctic cooling is also corroborated by the fact that in the 1990s the sign of the trend of the high-latitude zonality index characterizing the mean difference in the elevation of the AT-500 surface between $60°$N and $80°$N changed from plus to minus, and the recurrence trend of the Arctic High became positive (Dmitriev, 2007). These variations point to a climate change turning point manifested as a start of filling of the Arctic circumpolar vortex. The consequences of this process will become clear in the coming decades.

7

Conclusions

Regular airborne and satellite observations, supplemented by reconstructions using historical data of shipboard observations, have allowed us to construct a long series of values of ice cover area in Arctic seas in the region from the eastern coasts of Greenland to Alaska. These data have been used for investigations of climatic variations at an inter-decadal scale in the Arctic.

1   Long-term (climatic) variations of ice cover state in the Arctic Ocean not only exert significant influence on economic activity of states adjacent to this region, but also affect the Earth's climate more broadly through feedback mechanisms.

2   The variations of ice cover in the seas of the Eurasian shelf were spectrally analyzed, revealing the presence of cycles with time periods of 50–60 years, about 20 years, 8–12 years, 5–7 years, and 2–3 years. The 50–60 year cycle characterizes the epochs of rise and fall of air temperature in the Arctic. In the western region (the Greenland, Barents, Kara seas), long-term cycles prevail. The shorter cycles are typical for the eastern region (the Laptev, East Siberian, Chukchi seas). These cyclic oscillations of sea ice extent were superimposed on the background consisting of a negative long-term linear trend that characterizes gradual decrease of sea ice extent during the 20th century and the beginning of the 21st century. It is conjectured that this apparent linear trend may be a segment of a super-secular (200 years) climatic cycle.

3   Significant variations of ice cover thickness and concentration correspond to cyclic variations in air temperature. The regional distribution of maximum variations of ice thickness depends on the location of boundaries of residual ice and the dynamics of ice cover. Climatic variations of ice thickness are not so noticeable in land-fast ice, as they are mainly influenced by thermodynamic processes.

4 Climatic variations of ice cover extent in Arctic seas are conjugated with variations of other hydro-meteorological factors; large-scale air temperature and atmospheric pressure fields are the most important ones. The temporal structure of their multi-annual variability is characterized by the same features as for ice extent variability.

5 Corresponding anomalies of atmospheric circulation, related to evolution of polar eddies, are an important factor in causing variability of various climatic indices in high and moderate latitudes of the Earth at the inter-decadal scale. The North polar eddy intensifies during warm epochs and partly fills up during cold periods. These variations have an effect on the state of the Arctic anticyclone, as well as on intensity and location of the belt of zonal transportation in the atmosphere.

6 Variations of general atmospheric circulation appear to be responsible for the 50–60 year cyclic oscillations in surface air temperature, the displacement of its maximum anomalies from high to moderate latitudes, and the opposite signs of anomalies over continents and oceans. These may be related to the conjectured presence of longer (200-year) cycles.

7 The scheme of general ice drift in the Arctic Basin varies in accordance with oscillations of baric fields. The cyclonic component of ice cover drift intensifies during the warm epochs, as compared with cold ones. During warm periods, ice outflow to the Greenland Sea weakens, the temperature of the deep Atlantic waters rises, the salinity of the surface waters increases, and river runoff volume into Arctic seas increases. Changes in ice drift tend to determine the displacement of the boundary of multi-annual ice towards the seas of the Eurasian shelf during the warm epochs, though relatively short-term local variations of wind fields can create significant departure from this regularity.

8 The climatic variations of the extent of ice cover and other hydro-meteorological characteristics are caused by processes in the atmosphere and ocean, that in turn are affected both by external and internal factors.

9 The most significant external factor is variability of the total solar irradiance (TSI). This includes variations of "usual" electromagnetic solar radiation (in the visible and infrared frequency ranges), as well as variability of solar activity (SA) due to processes within the Sun that produce variations in ultraviolet energetic particle fluxes and the magnetic field of the Sun.

10 SA variations have been characterized by cycles with periods of about 10, 20, 60 and 200 years. These cycles are likely to be related to changes in the Arctic climate. Despite the relatively small variations of TSI that may be associated with variations in SA, variability of SA has a significant effect on high latitude regions because the interaction of charged particles of the solar wind with the Earth's magnetic field concentrates these particles at high latitudes. Variability of conventional radiation has an effect on both high and moderate latitudes, where

there are seasonal variations in the interaction between the atmosphere and underlying surface during the course of a year.

11   The magnitude of solar radiation depends on the square of the Sun–Earth distance. Most previous models for TSI did not consider the phenomenon of "dissymmetry" of the solar system (the distance between the Sun and the center of gravity of the solar system). This distance varies under the influence of the greatest planets Jupiter and Saturn with a period close to 60 years. It appears possible that the observed 60-year cycle in Arctic climate is related to this variation.

12   While many climatologists have focused on greenhouse gases as the likely cause of global warming in the 20th century, an alternative explanation is that changes in solar irradiance (TSI) made significant contributions to climate change in the 20th century. The Hoyt and Schatten model for TSI follows a similar trend to that of Arctic temperatures in the 20th century. A SA reconstruction of sunspot numbers over a 1,200-year time period shows some similarity to reconstructions of past surface temperature from proxies. Over this time period it seems possible that solar variations were the prime factor in climate changes.

13   The Earth's climate is affected by internal and external factors. The internal factors include natural hydro-meteorological, geological, and biological processes, as well as self-oscillation phenomena related to interactions in the ocean-sea ice-atmosphere-glaciers system. In addition, anthropogenic impacts are also considered to be internal factors; they are caused by the increase in concentration of greenhouse gases in the atmosphere because of human activity. External factors include solar activity, tidal and nutation phenomena, variability of the Earth's rotation speed, fluctuations in the solar constant, fluxes of energy and charged particles from space, and other astronomical factors.

14   Many climatologists have concluded that anthropogenic factors burning of fossil fuels, deforestation and other processes exerted a strong influence on global warming in the 20th century. This was based on coupled-model simulations of the general circulation of the ocean, atmosphere, and ice cover. However, these models do not appear to reflect the cyclic features of variations in Arctic ice extent and climate.

15   Assuming that our cyclic interpretation of 20th century variations in Arctic climate is correct, recurring features in air temperature and ice extent allow us to extrapolate the cycles forward into the twenty-first century. According to these forecasts, continuing natural cyclic changes will bring about both decreases and increases in the ice extent of Arctic Ocean marginal seas. Based on ice conditions expected during the first half of the twenty-first century, there will likely be a continuing need for icebreaking support of marine operations in the Arctic.

Appendix A

Mean monthly ice index values in April and August for the Eurasian Arctic Seas for 1900–2008

(1) Greenland Sea (April). *(2)* Barents Sea (April). *(3)* Greenland Sea (August).
(4) Barents Sea (August). *(5)* Kara Sea (August). *(6)* Laptev Sea (August).
(7) East-Siberian Sea (August). *(8)* Chukchi Sea (August).

Year	*(1)*	*(2)*	*(3)*	*(4)*	*(5)*	*(6)*	*(7)*	*(8)*
1900	663.07	1005.12	597.85	319.00	473.00	279.00	578.00	100.44
1901	804.38	1016.86	467.41	319.00	540.00	284.08	554.00	186.00
1902	804.38	1218.20	358.71	389.00	423.00	428.80	551.79	111.60
1903	793.51	941.92	326.10	574.00	664.00	209.04	570.00	122.76
1904	684.81	792.03	402.19	291.00	457.00	370.00	662.00	134.00
1905	913.08	792.03	423.93	264.00	581.00	236.00	546.00	149.00
1906	630.46	904.90	315.23	208.00	498.00	354.00	693.00	201.00
1907	978.30	816.41	423.93	264.00	581.00	251.64	539.00	93.00
1908	989.17	916.63	315.23	208.00	423.00	230.00	525.00	150.81
1909	760.90	1016.86	358.71	430.00	415.00	301.97	693.00	186.00
1910	739.16	954.56	391.32	486.00	376.13	226.48	732.00	201.00
1911	847.86	967.20	347.84	361.00	540.00	226.48	539.00	74.00
1912	913.08	1192.92	543.50	574.00	747.00	428.80	677.60	201.00

Year	(1)	(2)	(3)	(4)	(5)	(6)	(7)	(8)
1913	760.90	1042.14	445.67	444.00	581.00	160.80	539.00	167.00
1914	869.60	979.84	380.45	458.00	664.00	214.40	693.00	149.00
1915	597.85	941.92	456.54	416.00	374.00	226.48	426.38	149.00
1916	793.51	1180.28	576.11	361.00	581.00	354.00	539.00	74.00
1917	858.73	1242.58	532.63	583.00	664.00	204.00	500.00	93.00
1918	858.73	1092.70	445.67	472.00	564.40	204.00	501.63	104.16
1919	858.73	941.92	326.10	305.00	498.00	402.00	693.00	111.60
1920	793.51	766.75	369.58	305.00	415.00	279.00	616.00	186.00
1921	728.29	766.75	336.97	347.00	664.00	251.64	677.20	130.00
1922	804.38	841.69	239.14	153.00	423.00	226.48	601.95 1	30.20
1923	804.38	841.69	434.80	125.00	415.00	327.14	576.87	200.88
1924	673.94	816.41	347.84	208.00	538.00	413.00	747.00	193.00
1925	521.76	804.68	315.23	222.00	687.80	305.30	628.30	190.00
1926	673.94	979.84	456.54	333.00	796.80	289.50	619.30	82.10
1927	717.42	1016.86	434.80	333.00	610.90	347.40	697.30	141.70
1928	847.86	888.19	402.19	277.56	659.70	510.60	736.20	223.50
1929	826.12	1235.14	445.67	388.58	653.20	305.30	736.20	204.90
1930	467.41	1013.09	250.01	111.02	548.60	295.30	736.70	212.30
1931	597.85	943.70	347.84	97.15	423.20	331.60	743.90	230.90
1932	728.29	1040.85	293.49	194.29	182.50	205.30	713.60	212.30
1933	456.54	804.92	228.27	55.51	553.90	449.80	649.20	197.50
1934	673.94	846.56	445.67	138.78	671.50	210.90	575.90	156.50
1935	608.72	1068.61	358.71	277.56	399.80	276.80	669.10	123.10
1936	739.16	999.22	163.05	111.02	572.30	357.80	696.30	186.30
1937	673.94	971.46	347.84	27.76	357.40	342.30	483.50	123.10
1938	673.94	846.56	336.97	55.51	162.00	260.10	670.00	167.70
1939	500.02	888.19	336.97	55.51	474.60	287.80	618.50	119.30

Year	(1)	(2)	(3)	(4)	(5)	(6)	(7)	(8)
1940	663.07	874.31	347.84	235.93	252.60	292.30	593.30	71.00
1941	706.55	971.46	391.32	124.90	168.00	289.40	706.30	130.50
1942	641.33	1207.39	326.10	124.90	287.80	263.00	741.20	201.20
1943	728.29	1054.73	434.80	194.29	262.70	141.40	544.50	74.70
1944	684.81	763.29	369.58	194.29	295.60	378.90	607.60	111.90
1945	641.33	902.07	326.10	208.17	94.60	93.90	583.30	164.00
1946	858.73	860.44	282.62	124.90	434.60	303.80	645.50	160.30
1947	706.55	763.29	304.36	152.66	332.80	202.50	621.40	171.40
1948	804.38	721.66	391.32	152.66	374.10	370.40	654.30	123.10
1949	706.55	902.07	391.32	152.66	461.60	440.40	679.30	93.30
1950	641.33	846.56	315.23	124.90	352.30	193.80	661.10	59.80
1951	782.64	804.92	304.36	222.05	289.20	273.40	584.80	74.70
1952	771.77	971.46	467.41	138.78	399.20	201.80	542.10	108.20
1953	760.90	777.17	369.58	27.76	356.90	122.70	474.20	104.50
1954	967.43	777.17	206.53	27.76	252.70	259.90	664.10	145.40
1955	641.33	693.90	271.75	27.76	124.40	295.50	673.20	212.30
1956	586.98	846.56	282.62	83.27	409.90	411.30	578.90	208.60
1957	336.97	971.46	260.88	83.27	368.50	460.70	522.40	186.30
1958	347.84	1026.97	260.88	194.29	603.70	319.90	578.20	111.90
1959	554.37	971.46	347.84	208.17	277.70	165.10	566.80	93.30
1960	467.41	763.29	250.01	69.39	367.30	143.80	485.70	89.60
1961	630.46	846.56	293.49	166.54	221.40	226.10	630.20	119.30
1962	565.24	943.70	347.84	277.56	473.50	484.80	708.60	89.60
1963	641.33	1151.87	391.32	333.07	543.80	466.90	736.20	119.30
1964	684.81	943.70	326.10	277.56	607.70	265.30	538.70	119.30
1965	728.29	832.68	423.93	152.66	296.40	205.40	733.50	186.30
1966	521.76	1193.51	336.97	263.68	763.40	323.30	713.10	93.30

Year	(1)	(2)	(3)	(4)	(5)	(6)	(7)	(8)
1967	663.07	763.29	456.54	249.80	519.50	345.60	721.70	82.10
1968	826.12	888.19	576.11	263.68	547.90	97.00	368.60	37.50
1969	858.73	1096.36	358.71	388.58	700.10	321.40	572.00	149.10
1970	739.16	804.92	347.84	111.02	581.80	269.10	649.60	156.50
1971	684.81	832.68	369.58	124.90	453.50	182.00	601.50	149.10
1972	663.07	902.07	315.23	41.63	467.10	459.00	710.40	74.70
1973	619.59	721.66	380.45	83.27	496.60	155.60	638.80	167.70
1974	510.89	777.17	293.49	138.78	639.50	269.50	483.80	97.00
1975	652.20	763.29	434.80	166.54	268.40	326.50	736.20	201.20
1976	434.80	680.02	347.84	83.27	285.90	363.30	575.60	175.10
1977	673.94	874.31	380.45	166.54	427.80	148.70	513.40	93.30
1978	652.20	971.46	336.97	249.80	462.90	236.40	657.40	89.60
1979	641.33	1124.12	336.97	222.05	383.30	412.80	751.20	108.20
1980	467.41	860.44	347.84	180.41	571.80	235.60	703.10	134.20
1981	467.41	888.19	423.93	194.29	599.60	296.30	572.00	156.50
1982	554.37	1013.09	369.58	263.68	416.50	365.60	710.80	119.30
1983	434.80	680.02	326.10	111.02	423.20	201.70	670.00	164.00
1984	369.58	749.41	347.84	13.88	308.90	414.90	746.70	175.10
1985	576.11	929.83	260.88	83.27	276.30	320.40	732.60	111.90
1986	706.55	957.58	315.23	124.90	501.40	334.40	693.60	137.90
1987	510.89	1013.09	423.93	152.66	453.80	364.10	636.90	97.00
1988	684.81	915.95	358.71	111.02	436.30	247.00	671.30	141.70
1989	554.37	652.27	423.93	222.05	457.40	397.80	459.80	134.20
1990	456.54	610.63	326.10	97.15	460.80	109.80	79.40	33.80
1991	478.28	680.02	239.14	124.90	479.00	112.50	494.10	126.80
1992	521.76	693.90	391.32	97.15	379.80	282.50	455.40	104.50
1993	402.19	804.92	271.75	194.29	391.90	352.00	496.30	108.20

Year	(1)	(2)	(3)	(4)	(5)	(6)	(7)	(8)
1994	358.71	791.05	358.71	55.51	265.20	235.60	670.00	160.30
1995	402.19	346.95	369.58	13.88	44.80	44.50	457.80	108.20
1996	586.98	749.41	271.75	111.02	491.50	419.50	716.30	93.30
1997	630.46	971.46	413.06	83.27	233.30	228.90	653.30	30.10
1998	434.80	1026.97	326.10	152.66	450.50	384.90	710.80	204.90
1999	380.45	915.95	250.01	111.02	551.80	70.60	656.50	71.00
2000	521.76	680.02	358.71	13.88	178.40	117.90	590.30	137.90
2001	576.11	777.17	217.40	13.88	251.50	314.00	690.30	107.88
2002	369.58	666.14	119.57	83.27	347.30	204.90	339.00	89.28
2003	500.02	846.56	195.66	152.66	470.85	213.30	385.11	70.68
2004	347.84	707.78	130.44	55.51	244.75	382.33	624.03	100.44
2005	478.28	707.78	228.00	14.00	141.00	134.00	339.00	7.00
2006	369.58	499.61	293.00	14.00	148.00	59.00	428.00	74.00
2007	445.67	485.73	304.00	42.00	236.00	144.00	0.00	0.00
2008	456.54	582.88	196.00	56.00	166.00	315.00	383.00	60.00

Appendix B

Mean annual surface air temperature (SAT) in the zone from 70–85°N for 1900–2007

1900	0.0	1916	−0.7	1932	0.7	1948	−0.7	1964	−1.2
1901	−0.6	1917	−1.9	1933	0.4	1949	0.1	1965	−0.4
1902	−2.0	1918	−1.3	1934	0.9	1950	0.1	1966	−1.3
1903	−0.7	1919	−0.5	1935	0.5	1951	0.0	1967	0.1
1904	−0.4	1920	−0.1	1936	0.6	1952	0.6	1968	−0.9
1905	−1.0	1921	−0.4	1937	1.1	1953	0.8	1969	−0.5
1906	−0.6	1922	0.2	1938	2.0	1954	1.0	1970	0.0
1907	−0.3	1923	0.5	1939	0.7	1955	−0.5	1971	−0.1
1908	0.1	1924	0.0	1940	1.0	1956	0.3	1972	0.1
1909	−0.3	1925	−0.4	1941	−0.2	1957	−0.2	1973	−0.3
1910	−0.7	1926	0.0	1942	0.6	1958	−0.3	1974	0.0
1911	0.1	1927	0.2	1943	0.9	1959	0.4	1975	−0.3
1912	−0.9	1928	0.8	1944	1.0	1960	0.4	1976	−0.1
1913	−1.0	1929	0.1	1945	0.5	1961	−0.6	1977	−0.2
1914	−1.1	1930	1.1	1946	0.1	1962	−0.2	1978	−0.4
1915	−1.0	1931	0.9	1947	1.1	1963	−1.1	1979	−0.9

Year	SAT	Year	SAT	Year	SAT	Year	SAT	Year	SAT
1980	0.0	1986	−0.2	1992	−0.3	1998	1.7	2004	0.7
1981	0.9	1987	−0.6	1993	0.3	1999	0.7	2005	2.0
1982	−0.3	1988	0.1	1994	0.7	2000	1.1	2006	2.1
1983	0.1	1989	−0.1	1995	2.0	2001	0.4	2007	1.9
1984	0.8	1990	0.8	1996	1.3	2002	1.2		
1985	0.5	1991	0.8	1997	1.1	2003	1.4		

Appendix C

Mean annual zonality index in the atmosphere of the zone from 40–65°N for 1900–2007

1900	1.363	1916	0.483	1932	0.931	1948	2.237	1964	−0.667
1901	0.206	1917	−2.133	1933	−1.723	1949	1.371	1965	−1.721
1902	−0.623	1918	1.200	1934	1.000	1950	0.104	1966	−3.020
1903	−1.400	1919	−0.400	1935	−0.300	1951	−0.986	1967	1.346
1904	0.751	1920	1.900	1936	−1.432	1952	−0.844	1968	−2.711
1905	0.739	1921	0.600	1937	−0.858	1953	−0.028	1969	−2.582
1906	−2.431	1922	−1.074	1938	2.000	1954	−0.910	1970	−0.517
1907	−2.132	1923	1.886	1939	0.700	1955	−1.225	1971	−0.045
1908	0.147	1924	0.071	1940	−2.528	1956	−1.377	1972	−0.291
1909	−1.538	1925	1.120	1941	−1.823	1957	−0.646	1973	1.146
1910	0.105	1926	−0.959	1942	−1.057	1958	−2.286	1974	−1.047
1911	2.029	1927	−0.471	1943	1.570	1959	−0.838	1975	1.261
1912	0.587	1928	3.000	1944	−0.836	1960	−2.528	1976	0.350
1913	1.667	1929	0.495	1945	−1.500	1961	−0.502	1977	0.213
1914	2.278	1930	0.944	1946	1.300	1962	−1.141	1978	1.534
1915	−2.385	1931	0.560	1947	−2.100	1963	−2.633	1979	0.872

1980	−1.729	1986	1.346	1992	1.841	1998	1.441	2004	0.881
1981	−0.442	1987	0.137	1993	1.272	1999	2.208	2005	−0.600
1982	2.286	1988	1.280	1994	2.229	2000	1.290	2006	2.600
1983	0.631	1989	3.266	1995	−0.899	2001	0.798	2007	2.500
1984	0.014	1990	4.400	1996	−2.296	2002	1.367		
1985	−0.582	1991	1.056	1997	0.519	2003	1.270		

References

Aagaard, K., and E. C. Carmack (1989). The role of sea ice and other fresh water in the Arctic circulation. *J. Geophys. Res.*, **94**, 14485–14498.

Abramov, R. V. (1967). Some consequences of geographical detailing of the classic concept of the atmosphere action centers. *Proc. LGMI*, **24**, 22–30 [in Russian].

Abramov, V. A., and I. Ye. Frolov (1987). Spatial non-uniformity of the heat exchange of the Barents Sea with the atmosphere and mean monthly trajectories of cyclones in the wintertime. In *Abstracts of the Third Congress of Soviet Oceanographers, December 14–19: Physics and Chemistry of the Ocean Section, Polar and Regional Oceanography*. Leningrad: Gidrometeoizdat, pp. 123–124 [in Russian].

Aleksandrov, E. I., N. N. Bryazgin, A. A. Dementyev, and V. F. Radionov (2004). *Meteorological Regime of the Arctic Basin (Drifting-station Data)*, Vol. 11: *Climate of the Near-ice Atmospheric Layer of the Arctic Basin*. St. Petersburg: Gidrometeoizdat, 144 pp. [in Russian].

Alekseev, G. V. (1976). On the relation of the ice cover state and the atmosphere in the central Arctic. *AARI Proc.*, **332**, 109–113 [in Russian].

Alekseev, G. V. (1995). Interaction of atmosphere and ocean in the polar regions. *Problems of the Arctic and the Antarctic*, **70**, 193–2002 [in Russian].

Alekseev, G. V. (2000). Modern climate changes in the Arctic. *Problems of the Arctic and the Antarctic*, **72**, 42–71 [in Russian].

Alekseev, G. V. (2003). Studies of climate changes in the Arctic in the XX century. *AARI Proc.*, **446**, 6–21 [in Russian].

Alekseev, G. V., and N. E. Ivanov (2003). Regional and seasonal features of warming in the Arctic in the 1930's and 1990's. *AARI Proc.*, **446**, 41–47 [in Russian].

Alekseev, G. V., and P. N. Svyashchennikov (1991) *Natural Variability of Climate Characteristics of the Northern Polar Area and the Northern Hemisphere*. Leningrad: Gidrometeoizdat, 160 p. [in Russian].

Alekseev, G. V., O. I. Myakoshin, and N. P. Smirnov (1997). Variability of ice transport across Fram Strait. *Meteorology and Hydrology*, **9**, 12–17.

Alekseev, G. V., V. F. Bulatov, V. F. Zakharov, and V. V. Ivanov (1998a). Heat expansion of Atlantic water in the Arctic Basin. *Meteorology and Hydrology*, **7**, 69–78 [in Russian].

Alekseev, G. V., V. F. Zakharov, A. H. Smirnov, and N. P. Smirnov (1998b). Multiyear fluctuations of ice conditions and atmospheric circulation in the sub-Atlantic Arctic and North Atlantic. *Meteorology and Hydrology*, **9**, 87–98.

Alekseev, G. V., V. F. Bulatov, and V. F. Zakharov (2000). Role of the Arctic High in freshwater distribution in the Arctic Basin. *Meteorology and Hydrology*, **2**, 61–68.

Alekseev, G. V., S. I. Kuzmina, O. G. Aniskina, and N. E. Kharlanenkova (2003). Natural and anthropogenic constituents of changes in near-surface air temperature in the Arctic in the XX century based on observation and modeling data. *AARI Proc.*, **446**, 22–30.

Alexandrov, V. Y., T. Martin, J. Kolatschek, H. Eicken, M. Kreyscher, and A. P. Makshtas (2000). Sea ice circulation in the Laptev Sea and ice export to the Arctic Ocean: Results from satellite remote sensing and numerical modeling. *J. Geophys. Res.*, **105**(C7), 17143–17159.

Alekseev, G. V., E. I. Aleksandrov, R. V. Bekryaev, V. V. Ivanov, A. A. Korablev, V. V. Maistrova, A. P. Nagurny, V. F. Radionov, Zakharov V. F. *et al.* (2004). In G. V. Alekseev (Ed.), *Formation and Dynamics of Modern Climate of the Arctic Regions.* St. Petersburg: Gidrometeoizdat, 266 pp. [in Russian].

Anon. (A). Available at *http://www.theaustralian.news.com.au/story/0,25197,24365156-11949,00.html*

Anon. (B). Available at *http://sydney.indymedia.org.au/story/climate-change-arctic-sea-ice-heading-rapid-disintegration*

Anon. (C). Available at *http://www.encyclopedia.com/doc/1G1-168953938.html*

Anon. (D) (1966). *Atlas of the Antarctic.* Moscow: GU GK of the USSR MG [in Russian].

Anon. (E) (1985). *Atlas of the Arctic.* Moscow: GU GiK of the USSR CM [in Russian].

Anon. (F) (1980). *Atlas of the Oceans: The Arctic Ocean.* Moscow: GUNiO USSR, 190 pp.

Anon. (G) (1996). *Atlas of the Water Balance of the Northern Polar Area.* St. Petersburg: Gidrometeoizdat, 82 pp. [in Russian].

Anon. (H) (1965). *Instruction on Assessing the Quality of the Methods and Skill Score of Marine Hydrological Forecasts.* Moscow: Gidrometeoizdat, 88 pp. [in Russian].

Anon. (I) (1935). *Sailing Directions for the Kara Sea, Part 2.* GO UMS RKK and GU of Glavsevmorput', 429 pp. [in Russian].

Anon. (J) (1938). *Sailing Directions for the Laptev Sea.* GU RKKF and GU of Glavsevmorput', 202 pp. [in Russian].

Anon. (K) (1938). *Sailing Directions for the Chukchi Sea.* GU RKKF and GU of Glavsevmorput', 159 pp. [in Russian].

Anon. (L) (1939). *Sailing Directions for the East-Siberian Sea.* Hydrographic Administration of Glavsevmorput', 124 pp. [in Russian].

Anon. (M) (1970). *WMO Sea-Ice Nomenclature: Terminology, Codes and Illustrated Glossary,* WMO/TD-259. Geneva: World Meteorological Organization, 147 pp.

Anon. (N) (1974). *World Water Balance and Water Resources of the Earth.* Leningrad: Gidrometeoizdat, 638 pp. [in Russian].

Anon. (O) (1960). *Boundaries of the Oceans and the Seas.* Administration of the Head of the GS VMF, 52 pp. [in Russian].

Appel, I. L., and Z. M. Gudkovich (1984). Study of possible changes of average salinity of the upper layer of the Kara Sea caused by stable anomalies of the upper layer of the river runoff. *Problems of the Arctic and the Antarctic*, **58**, 5–14.

Appel, I. L., and Z. M. Gudkovich (1992). *Numerical Modeling and Forecasting of the Ice Cover Evolution in the Arctic Seas during the Period of Melting.* St. Petersburg: Gidrometeoizdat, 144 pp. [in Russian].

ACIA (Arctic Climate Impact Assessment) (2005). New York: Cambridge University Press. 1042 pp.

Arctic Climatology Project Environmental Working Group (2000). In F. Tanis and V. Smolyanitsky (Eds.), *Joint U.S.–Russian Sea Ice Atlas.* Boulder, CO: National Snow and Ice Data Center, CD-ROM.

Asmus, V. V., V. A. Krovotyntsev, O. E. Milekhin, and I. R. Trenina (2005). *Study of Multiyear Dynamics of Sea Ice in the Arctic from Satellite Radar Data: Issues of Processing and Interpretation of the Earth's Radar Sounding Data.* St. Petersburg: Gidrometeoizdat [*Proc. RC Planeta*, **1**(46), 155–172] [in Russian].

Bagrov, N. A. (1959). Analytical presentation of the sequences of meteorological patterns by means of the empirical orthogonal function. *TSIP Proceedings*, **74**, 3–24.

Baidal, M. Kh. (2001). *Main Climate Features of Kaluga and Adjacent Areas in 2001–2025.* Obninsk, 92 pp. [in Russian].

Baranov, G. I., and T. G. Vangengeim (1988). Natural orthogonal functions of the winter fields of a complex of climatic characteristics of the northern hemisphere. *AARI Proc.,* **404,** 36–46.

Baranov, G. I., V. K. Kurazhov, and R. A. Cheikina (1986). Diagnosis of the evolution of large-scale fluctuations of atmospheric circulation in the middle troposphere by means of natural orthogonal components. *AARI Proc.,* **393,** 103–109.

Bashkirtsev, V. R., and G. P. Mashnich (2004). Variability of the Sun and climate of the Earth. *Solar–Terrestrial Physics,* **6,** 135–137.

Belkin, I. M., S. Levitus, J. Antonov, and S-A. Malmberg (1998). Great salinity anomalies in the North Atlantic. *Progress in Oceanography,* **41,** 1–68.

Blinova, E. N. (1943). Hydrodynamic theory of air pressure waves, temperature waves and centers of atmospheric forcing. *Russian Academy of Sciences Reports,* **39**(7), 284–287.

Borisenkov, Ye. P. (1982). *Climate and Human Activity.* Moscow: Nauka, 136 pp. [in Russian].

Borisenkov Ye. P., and V. M. Pasetsky (2003). *Annals of Unusual Natural Phenomena for 2.5 Millennia (5th Century BC–20th Century AD).* St. Petersburg: Gidrometeoizdat, 536 pp. [in Russian].

Borodachev, V. E., and V. I. Shilnikov (2002). *History of the Ice Air Reconnaissance in the Arctic and the Russian Ice Covered Seas during 1914–1993.* St. Petersburg: Gidrometeoizdat, 444 pp. [in Russian].

Bothmer, V., and R. Schwenn (1998). The structure and origin of magnetic clouds in the solar wind. *Ann. Geophysicae,* **16,** 1–24.

Brooks, K. (1952). *Climates of the Past.* Moscow: IL Publishers, 358 pp. [in Russian].

Budyko, M. I. (1962). Some ways of impact on climate. *Meteorology and Hydrology,* **2,** 3–8.

Budyko, M. I. (1966). *Possible Climate Changes at Impact on Polar Ice: Modern Problems of Climatology.* Leningrad: Gidrometeoizdat, 227 pp. [in Russian].

Budyko, M. I. (1968). On the origin of glacial epochs. *Meteorology and Hydrology,* **11,** 3–12.

Budyko, M. I. (1969). *Polar Ice and Climate.* Leningrad: Gidrometeoizdat, 36 pp. [in Russian].

Budyko, M. I. (1982). Anthropogenic global climate change. *Vestnik AN SSSR,* **5,** 91–94.

Buinitsky, V. Kh. (1951). The formation and drift of the ice cover in the Arctic Basin. *Proceedings of the Drifting Expedition of Glavsevmorput onboard the Icebreaking Ship G. Sedov in 1937–1940,* Vol. 4, pp. 74–179.

Bulatov, L. V., and V. F. Zakharov (1978). On the change of the thermal state of the Arctic Ocean. *AARI Proc.,* **349,** 26–33.

Bulgakov, N. P. (1975). *Convection in the Ocean.* Moscow: Nauka, 272 pp. [in Russian].

Buzin, I. V. (2006). Estimations of some components of ice conditions in the northeastern Barents Sea. *International Journal of Offshore and Polar Engineering,* **16**(4), 274–282.

Buzuyev, A. Ya., and V. F. Dubovtsev (2002). Generalization of ice cover characteristics for assessing climatic changes in the Arctic Basin and the Siberian Shelf Seas. *Scientific Conference of CIS Countries.* St. Petersburg: Gidrometeoizdat, pp. 222–223 [in Russian].

Buzuyev, A. Ya., Yu. A. Gorbunov, Z. M. Gudkovich, S. M. Losev, and Ye. U. Mironov (1999). Study of morphometry and dynamics of the ice cover of the Arctic Basin. *Problems of the Arctic and the Antarctic,* **71,** 106–128.

Chaplygin, Ye. I., and A. V. Yanes (1968). Cosmic and global factors in the problem of background oceanographic forecasts. *AARI Proc.,* **285,** 233–238.

Divine, D. V., and C. Dick (2006). Historical variability of sea ice edge position in the Nordic Seas. *J. Geophys. Res.,* **111**(C01001), 1–14.

Dmitriev A. A. (1994). *Variability of Atmospheric Processes in the Arctic and Their Role in Long-Range Forecasting.* Leningrad: Gidrometeoizdat, 207 pp. [in Russian].

Dmitriev, A. A. (2000). *Dynamics of the Atmospheric Processes over the Seas of the Russian Arctic.* St. Petersburg: Gidrometeoizdat, 234 pp. [in Russian].

Dmitriev, A. A. (2007). On the causes of a natural phenomenon in the Arctic in the summer of 2007. *Problems of the Arctic and the Antarctic,* **77,** 115–127.

Dmitriev, A. A., and V. A. Belyazo (2006). *Outer Space, Planetary Climatic Variability and Atmosphere of Polar Regions.* St. Petersburg: Gidrometeoizdat, 360 pp. [in Russian].

Dobrovolsky, S. (2002). *Climatic Changes in the Hydrosphere–Atmosphere System.* Moscow: GEOS, 232 pp. [in Russian].

Dobrovolsky, S. G. (2000). *Stochastic Climate Theory.* Heidelberg: Springer-Verlag, 296 pp.

Dolgin, I. M. (1968). *Climate of Free Atmosphere in the Soviet Arctic.* Leningrad: Gidrometeoizdat, 398 pp. [in Russian].

Doronin, Yu. P. (1959). On the thermal transformation of the lower layer of the atmosphere in the Arctic. *AARI Proc.,* **226,** 76–98.

Doronin, Yu. P. (1968). On the problem of destruction of Arctic ice. *Problems of the Arctic and the Antarctic,* **28,** 21–28.

Doronin, Yu. P. (1969). *Thermal Interaction of the Atmosphere and Hydrosphere in the Arctic.* Leningrad: Gidrometeoizdat, 300 pp. [in Russian].

Doronin, Yu. P., and D. Ye. Kheisin (1975). *Sea Ice.* Leningrad: Gidrometeoizdat, 317 pp. [in Russian].

Douglass, D. H., and B. D. Clader (2002). Climate sensitivity of the Earth to solar irradiance. *Geophysical Research Letters,* **29,** 33-1–33-4.

Dukhovskoy, D. S., M. A. Johnson, and A. Y. Proshutinsky (2004). Arctic decadal variability: An auto-oscillatory system of heat and fresh water exchange. *Geophysical Research Letters,* **31**(L03302), 1–4.

Duplessy, J. C. (1980). Isotopic studies. In J. Gribbin (Ed., translated from English), *Climate Changes.* Leningrad: Gidrometeoizdat, pp. 70–101 [in Russian].

Flohn, H. (1980). Bases of the geophysical glaciation model. In J. Gribbin (Ed.), *Climate Changes.* Leningrad: Gidrometeoizdat, pp. 331–356 [in Russian].

Frolich, C., and J. Lean (1998). The Sun's total irradiance: Cycles, trends and related climate change uncertainties since 1976. *Geophysical Research Letters,* **25.** 4377–4380.

Frolov, I. E. (Ed.) (2008). *Review of Hydrometeorological Processes in the Arctic Ocean in 2007.* St. Petersburg: AARI, 80 pp. Available at *http://www.aari.ru/resources/m0035/gm_review_2007.pdf*

Frolov, I. E. (Ed.) (2009). *Review of Hydrometeorological Processes in the Arctic Ocean in 2008.* St. Petersburg: AARI, 125 pp. Available at *http://www.aari.ru/resources/m0035/gm_review_2008.pdf* [in Russian].

Frolov, I. E., Z. M. Gudkovich, V. P. Karklin, and V. M. Smolyanitsky (2009, in press). 60-year cyclicity in climate changes in polar regions. *Materials of Glaciological Studies.*

Frolov, I., Z. M. Gudkovich V. F. Radionov, L. A. Timokhov, and A. V. Shirochkov (2005). *Scientific Studies in the Arctic,* Vol. 1: *Drifting Scientific "North Pole" Stations.* St. Petersburg: Nauka, 268 pp. [in Russian].

Gasyukov, P. R., and N. P. Smirnov (1967) Fluctuations of the baric field of the northern hemisphere in an 11-year cycle of solar activity. *Reports of the USSR Academy of Science,* **173**(3), 567–569.

German, G. R., and R. A. Goldberg (1981). *Sun, Weather and Climate.* Leningrad: Gidrometeoizdat, 320 pp. [in Russian].

Girs, A. A. (1960). *Bases of Long-Range Weather Forecasting.* Leningrad: Gidrometeoizdat, 560 pp. [in Russian].

Golubev, V. N., S. A. Sokratov, G. A. Rhzanitsyn, and A. V. Shashkov (2004). The role of congelative ice in the gas exchange of the surface geospheres. Diminishing glaciosphere: Facts and analysis. *Proceedings of the XIII Glaciological Symposium, St. Petersburg, May 24–28,* pp. 59–60 [in Russian].

Gorbunov, Yu. A., Yu. D. Bychenkov, S. M. Losev, I. Yu. Kulakov, and A. V. Provorkin (1985). On ice exchange through Fram Strait in 1979–1980. *AARI Proc.,* **396,** 101–109.

Gordienko, P. A., and D. B. Karelin (1945). Problems of ice motion and spreading in the Arctic Basin. *Problems of the Arctic and the Antarctic,* **3,** 5–35.

Gribbin, J., and H. H. Lamb (1978). Climatic change in historical times. In J. Gribbin (Ed.), *Climate Change.* Cambridge, U.K.: Cambridge University Press, pp. 68–82.

Groverman, B. S., and H. E. Landsberg (1979). *Reconstruction of Northern Hemisphere Temperature: 1579–1880*, Meteorology Program Publications No. 79-181 and 70182. College Park, MD: University of Maryland.

Gudkovich,, Z. M. (1961a). On the question of the nature of the Pacific Ocean current in Bering Strait and causes of seasonal changes in its intensity. *Oceanology*, **1**(4): 608–612.

Gudkovich, Z. M. (1961b). Relation of the ice drift in the Arctic Basin with ice conditions in the Soviet Arctic Seas. *Proceedings of the Oceanography Commission of the USSR Academy of Science*, **11**, 13–20.

Gudkovich, Z. M., and Doronin, Yu. P. (2001). *Drift of Sea Ice*. St. Petersburg: Gidrometeoizdat, 112 pp. [in Russian].

Gudkovich, Z. M., and S. V. Klyachkin (2001). Sea ice cover deformations in the spatially non-homogeneous wind field. *16th International Symposium on Okhotsk Sea Ice, Mombetsu, Hokkaido, Japan, February 4–8*, Abstracts of Symposium, pp. 403–404.

Gudkovich, Z. M., and Ye. G. Kovalev (1967). On the influence of water circulation in the Arctic Basin on ice distribution in the eastern region of the Soviet Arctic. *AARI Proc.*, **116**, 7–20.

Gudkovich, Z. M., and Ye. G. Kovalev (1997). Conformity of large-scale processes in the atmosphere, ocean and ice cover of the northern polar area. *AARI Proc.*, **437**, 17–29 [in Russian].

Gudkovich, Z. M., and Ye. G. Kovalev (2002a). On some mechanisms of cyclic climate changes in the Arctic and the Antarctic. *Oceanology*, **42**(6), 1–7 [in Russian].

Gudkovich, Z. M., and Ye. G. Kovalev (2002b). Fluctuations of sea-ice extent of the Russian Arctic Seas in the 20th century and assessment of its possible changes in the 21st century. Paper presented at *Hydrometeorological Support for Economic Activity in the Arctic and the Ice-covered Seas, St. Petersburg, March 27–29*, pp. 36–45.

Gudkovich, Z. M., and A. Ya. Nikolayeva (1963). Ice drift in the Arctic Basin and its relation to sea-ice extent of the Arctic Seas. *AARI Proc.*, **104**, 212 [in Russian].

Gudkovich, Z. M., and S. P. Pozdnyshev (1995). Seasonal and spatial changes in mean speeds of the ice drift and gradient currents in the East Greenland ice flow. *Problems of the Arctic and the Antarctic*, **69**, 116–123 [in Russian].

Gudkovich, Z. M., and V. F. Zakharov (1998). Role of marginal dynamic processes in the change of ice concentration in the Arctic Seas in summer. *Meteorology and Hydrology*, **3**, 65–71 [in Russian].

Gudkovich, Z. M., E. I. Saruhanyan, and N. P. Smirnov (1970a). Baric "near-pole tide" and its influence on the sea-ice extent of the Arctic Seas. *Oceanology*, **10**(3), 426–437 [in Russian].

Gudkovich, Z. M., E. I. Saruhanyan, and N. P. Smirnov (1970b). The "pole tide" in the atmosphere of high latitudes and the sea-ice extent fluctuations in the Arctic Seas. *Doklady AN SSSR*, **190**(4), 954–957 [in Russian].

Gudkovich, Z. M., A. A. Kirillov, Ye. G. Kovalev, A. V. Smetannikova, and V. A. Spichkin (1972). *Long-range Ice Forecasting Methodology for the Arctic Seas*. Leningrad: Gidrometeoizdat, 348 pp. [in Russian].

Gudkovich, Z. M., V. P. Karklin, I. F. Romantsova, and K. A. Teitelbaum (1981). Application of statistical analysis for assessing the dependence of ice conditions in the Kara Sea on river runoff. *AARI Proc.*, **384**, 53–60 [in Russian].

Gudkovich,, Z. M., V. F. Zakharov, E. O. Aksenov, and S. P. Pozdnyshev (1994). Interaction of modern climatic changes of atmosphere, ocean and ice cover in the Arctic. *Conference on the Dynamics of the Arctic Climate System, November 7–10, Göteborg*, Abstracts, p. A-9.

Gudkovich, Z. M., V. F. Zakharov, Ye. O. Aksenov, and S. P. Pozdnyshev (1997). Interrelation of modern climatic changes in the atmosphere, ocean and ice cover of the Arctic. *AARI Proc.*, **437**, 7–16 [in Russian].

Gudkovich, Z., L. Timokhchov, A. Proshutinsky, A. Koltyshev, and A. Garmanov (2003). Climatic salinity changes in the surface layer of the Arctic Ocean: Multi-year and seasonal variability of the Arctic Ocean freshwater budget. *Final Science Conference: Progress in Understanding the Arctic Climate System: The ACSYS Decade and Beyond, St. Petersburg,*

Russia, November 11–14, WCRP-118 (CD), WMO/TD No. 1232, World Climate Research Programme (WCRP) Arctic Climate System Study (ACSYS). Geneva: World Meteorological Organization.

Gudkovich, Z. M., Ye. G. Kovalev, and E. G. Nikiforov (2004). On the relation of the Earth's angular rotation speed to climate changes. *Izvestiya RGO*, **6**, 1–10 [in Russian].

Gudkovich, Z. M., V. P. Karklin, and I. Ye. Frolov (2005). Intra-secular changes in climate and sea ice area in the Eurasian Arctic Seas and their possible causes. *Meteorology and Hydrology*, **2005**(6), 5–14 [in Russian].

Gudkovich, Z. M., R. B. Guzenko, and V. P. Karklin (2007). On the question of climate variability of the general schema of the ice drift in the Arctic Basin. *Data from Glaciological Studies*, **102**, 18–191 [in Russian].

Gudkovich, Z. M., V. P. Karklin, E. G. Kovalev, V. M. Smolyanitsky, and I. E. Frolov (2008). Variations of sea ice cover and other constituents of the climate system in the Arctic and Antarctic in connection with polar vortex evolution. *Problems of the Arctic and the Antarctic*, **78**, 48–57 [in Russian].

Harder, M., M. Hilmer, and P. Lemke (1998). Simulated sea ice transport through the Fram Strait. *ACSYS Arctic Forecast*, **3**, 1028–1114.

Hassol, S. J. (2004). *Impact of Warming in the Arctic*. Cambridge, U.K.: Cambridge University Press, 140 pp.

Hoyt, D. V., and K. H. Schatten (1993). A discussion of plausible solar irradiance variations, 1700–1992. *J. Geophys. Res.*, **98**(A11), 18895–18906.

IICWG (International Ice Charting Working Group) (2008). *Ninth Meeting, October 20–24, Science Workshop: Global View of the Present Ice Conditions*. Available at *http://www. nsidc.org/noaa/iicwg/meetings.html*

IPCC (2001). Contribution of Working Group I to the Third Assessment Report of the Intergovernmental Panel on Climate Change. In J. T. Houghton, Y. Ding, D. J. Griggs, M. Noguer, P. J. van der Linden, X. Dai, K. Maskell, and C. A. Johnson (Eds.), *Climate Change 2001: The Scientific Basis*. Cambridge, U.K.: Cambridge University Press, 881 pp.

IPCC (2007). Contribution of Working Group I to the Fourth Assessment Report of the Intergovernmental Panel on Climate Change. In S. Solomon, D. Qin, M. Manning, Z. Chen, M. Marquis, K. B. Averyt, M. Tignor, and H. L. Miller (Eds.), *Climate Change 2007: The Physical Science Basis*. Cambridge, U.K.: Cambridge University Press, 996 pp.

Itin, V. (1933). *Sea Routes in the Soviet Arctic*. Moscow: Publishing House of Soviet Asia, 110 pp. [in Russian].

Ivanov, V. V. (1976). Freshwater balance of the Arctic Ocean. *AARI Proc.*, **323**, 138–147.

Ivanov, V. V. (1980). Hydrological regime of the low reaches and mouths of rivers in West Siberia and the problem of assessment of its changes under the influence of the territorial redistribution of water resources. *Problems of the Arctic and the Antarctic*, **55**, 20–43.

Ivanov, V. V., L. A. Timokhov, N. N. Bryazgin, Z. M. Gudkovich., S. V. Kochetov, and V. M. Smolyanitsky (2003). *Multi-Year and Seasonal Variability of the Arctic Ocean Freshwater Budget: ACSYS Final Science Conference, November 11–14*, Abstracts, p. 19. St. Petersburg: AARI of Roshydromet [in Russian].

Ivanov, V. V., O. V. Muzhdaba, and Z. R. Solovieva (2004). Multiyear changes in the annual and seasonal river water inflow to the Arctic Seas. *Sixth All-Russia Hydrological Congress, September 28–October 1*, Abstracts, pp. 39–41. St. Petersburg: Gidrometeoizdat [in Russian].

Izrael, Yu. A., V. Gruza, V. M. Katsov, and V. P. Meleshko (2001). Global climate changes: Role of anthropogenic impacts. *Meteorology and Hydrology*, **2001**(5), 5–21 [in Russian].

Johannessen, O. M., L. Bengtsson, M. W. Miles, S. I. Kuzmina, V. A. Semenov, G. V. Alekseev, A. P. Nagurnyi, V. G. Zakharov, L. P. Bobylev, L. H. Pettersson *et al.* (2004). Arctic climate change: Observed and modeled temperature and sea ice variability. *Tellus*, **56A**, 1–18.

Jose, P. D. (1965). Sun's motion and sunspots. *Astronomical Journal*, **70**(3), 193–200.

Karelin, D. B. (1951). Development of ice forecasting in connection with scientific observations onboard the icebreaking ship *G. Sedov*. In *Proceedings of the Drifting Expedition of Glavsevmorput' onboard the Icebreaking Ship G. Sedov in 1937–1940*, Vol. 4, pp. 180–195 [in Russian].

Karklin, V. P. (1973). Twenty-two year cycle of solar activity and the atmospheric pressure fields at temperate and high latitudes of the Northern Hemisphere. *Izvestiya VGO*, **1973**(3), 275–280 [in Russian].

Karklin, V. P. (1975). Role of solar activity in multiyear changes in the location and intensity of the Icelandic Low of atmospheric pressure. *Problems of the Arctic and the Antarctic*, **46**, 79–83 [in Russian].

Karklin, V. P. (1977). Quasi-biennial oscillations in sea-ice extent changes in the Arctic Seas. *AARI Proc.*, **341**, 103–113 [in Russian].

Karklin, V. P. (1978). Changes in the atmospheric pressure field at high and temperate latitudes of the northern hemisphere in 11-year cycles of solar activity. *Problems of the Arctic and the Antarctic*, **54**, 62–68 [in Russian].

Karklin, V. P. (1987). On the issue of the causes of 6–7-year oscillations of sea-ice extent of the Arctic Seas. *AARI Proc.*, **402**, 67–80 [in Russian].

Karklin, V. P., and Ye. G. Kovalev (1994). Influence of solar activity on the formation of large sea-ice extent anomalies of the Arctic Seas. *AARI Proc.*, **432**, 28–35 [in Russian].

Karklin, V. P., and K. A. Teitelbaum (1987). Temporal structure of multiyear changes in sea-ice extent in the Arctic Seas. *AARI Proc.*, **402**, 53–66 [in Russian].

Karklin, V. P., A. V. Yulin, I. D. Karelin, and V. V. Ivanov (2001). Climatic fluctuations of sea-ice extent of the Arctic Seas of the Siberian shelf. *AARI Proc.*, **443**, 5–11 [in Russian].

Katsov, V. M. (2003). Scenarios of climate changes in the Arctic in the 21st century. *Meteorology and Hydrology*, **10**, 5–19 [in Russian].

Keigwin, L. D. (1996). The Little Ice Age and Medieval Warm Period in the Sargasso Sea. *Science*, **274**, 1503–1508.

Khromov, R. P., and L. I. Mamontova (1974). *Meteorological Glossary*. Leningrad: Gidrometeoizdat, 568 pp. [in Russian].

Klimenko, V. V. (2007). Climatic sensation: What awaits us in the near and distant future? "Polit. ru" public lectures. Available at *www. polit.ru/lectures/2007/02/15/klimenko.html*

Klyashtorin, L. B., and A. A. Lyubushin (2003). On the coherence between the dynamics of world fuel consumption and global temperature anomaly. *Energy and Environment*, **14**(6), 773–782.

Klyashtorin, L. B., and A. A. Lyubushin (2004). Cyclic variability of climate and fisheries resources: Relevance for the Nordic Seas. In S. Skreslet (Ed.), *Jan Mayen Island in Scientific Focus: Proceedings of the NATO ARW on Joint International Scientific Observation Facility on Jan Mayen Island, November 11–15, 2003, Oslo, Norway* (NATO Science Series, IV: Earth and Environmental Sciences No. 45). Kluwer Academic.

Klyashtorin, L. B., and A. A. Lyubushin (2006). On the relationship of the global temperature anomaly and the world fuel consumption. *Modern Global Changes of the Environment*, Vol. 2. Moscow: Nauchnij mir, pp. 537–543 [in Russian].

Koesner, R. M. (1973). The mass balance of sea ice in the Arctic Ocean. *J. Glaciology*, **12**, 173–186.

Koltyshev, A. E., and L. A. Timokhov (1997). On the consistency of interannual sea-ice extent fluctuations of the Siberian shelf seas and surface circulation of the Arctic Basin. *AARI Proc.*, **437**, 173–192 [in Russian].

Kondratyev, K. Ya. (2004). Changes in global climate: Unsolved problems. *Meteorology and Hydrology*, **2004**(6), 118–127 [in Russian].

Kondratyev, K. Ya., and G. A. Nikolsky (1995). Solar activity and climate. *Study of the Earth from Space*, **1995**(5), 3–17 [in Russian].

Kovalenko, V. D., L. D. Kizim, and A. M. Pashestyuk (1987). Analysis of weather and climate variations. *Scientific Proceedings of SO VASHNIL*. Novosibirsk: SO VASKHNIL, 103 pp. [in Russian].

Korolev, V. K., and V. V. Subbotin (1988). Peculiarities of the structure of winter thermal baric fields of the northern hemisphere during the epochs of "warming" and "cooling" of the Arctic. *AARI Proc.*, **404**, 24–36 [in Russian].

Kovalev, Ye. G. (1960). Cyclicity in sea-ice extent fluctuations in the region of the New-Siberian Islands and the possibility of its use for forecasting. *Doklady AN SSSR*, **135**(2), 439–442 [in Russian].

Kovalev, Ye. G. (1967). Multiyear fluctuations of total sea-ice extent of the Arctic Seas. *AARI Proc.*, **116**, 21–35 [in Russian].

Kovalev, E G., and V. A. Spichkin (1977). Possibility of using a "method of superposition of the solar activity epochs" for long-range forecasting of sea-ice extent of the Arctic Seas. *AARI Proc.*, **346**, 89–93 [in Russian].

Kovalev, E. G., and A. V. Yulin (1998). Automated prognostic system for scientific-operational support of navigation in the Arctic. *AARI Proc.*, **438**, 73–82 [in Russian].

Krutskikh, B. A. (Ed.) (1991). *Climatic Regime of the Arctic at the Boundary of the 20th and 21st Centuries*. Leningrad: Gidrometeoizdat, 200 pp. [in Russian].

Krymsky, P. F. (1994). *A Possible Mechanism for Solar Wind Influence on Atmospheric and Geophysical Processes and the Earth's Rotation*. Yakutsk: Yakutsk Research Center of SO RAN, 64 pp. [in Russian].

Kupetsky, V. N. (1969). On the structure of helio-climatic relations and possibilities for their use in long-range and super-long-range forecasting. *Izvestiya VGO*, **4**, 289–295 [in Russian].

Kupetsky, V. N. (1974). Use of solar–terrestrial relations for long-range prediction of hydrometeorological phenomena. *Solar–Atmospheric Relations in Climate Theory and Weather Forecasting*. Leningrad: Gidrometeoizdat, pp. 452–462 [in Russian].

Kupetsky, V. N. (1977). On the use of geomagnetic perturbation for prediction of hydrometeorological phenomena. *AARI Proc.*, **340**, 138–143 [in Russian].

Kurazhov, V. K., V. A. Belyazo, A. A. Dmitriyev, and V. V. Ivanov (2004). Diagnosis of peculiarities of long-period fluctuations of atmospheric circulation and thermal regime in the Arctic in the 20th century and the forthcoming scenario of climate variability. *Express-Information, No. 18: Abstracts of the Concluding Session of the AARI Scientific Council*. St. Petersburg: AARI, p. 41 [in Russian].

Kuznetsov, A. P., and O. G. Sorokhtin (2000). *On the Greenhouse Effect: Global Environmental Changes*. Moscow: Scientific World, pp. 151–160 [in Russian].

Kwok, R., G. F. Cunningham, and S. S. Pang (2004). Fram Strait sea ice outflow. *J. Geophys. Res.*, **109**(C01009), doi: 10.1029/2003JC001785.

Lamb, H. H., and A. I. Johnson (1959; 1964—translation into Russian). Climatic variation and observed changes in the general circulation, Parts I and II. *Geografiska Annaler*, **41**, 94–134.

Lassen, K., and E. Friis-Christensen (1991). Similarity between long-term variations of polar sea-ice cover, global mean temperature and solar activity. Paper presented at *XVI EGS General Assembly, Wiesbaden, April 22–26*.

Latukhov, S. V., and B. A. Sleptsov-Shevlevich (1995). *Ice Conditions of Navigation in the Western Arctic*. St. Petersburg, 148 pp. [in Russian].

Laushkin, R. I. (1962). Calculation of the dynamic component of the ice balance of the Greenland Sea. *LGMI Proc.*, **16**, 70–74 [in Russian].

Lebedev, A. A., and N. R. Uralov (1977). On the issue of the ice balance of the Greenland Sea. *AARI Proc.*, **341**, 43–52 [in Russian].

Lesgaft, E. (1913). *Arctic Ocean Ice and a Seaway from Europe to Siberia*. St. Petersburg, 237 pp. [in Russian].

Lindzen, R. S. (1997). Can increasing carbon dioxide cause climate change? *Proceedings of the National Academy of Sciences of the U.S.A.*, **94**, 8335–8342.

Lineikin, P. R. (1955). On wind currents in the baroclinic layer of the sea. *GOIN Proc.*, **29**(41), 34–64 [in Russian].

Lipenkov, V. Ya., A. A. Yekaikin, and A. N. Salamatin (2002). Results of paleo-climatic studies from pits and ice core of the deep borehole at Vostok station. *Express-Information, No 12: Abstracts of the Concluding Session of the AARI Scientific Council*. St. Petersburg: AARI, pp. 93–94 [in Russian].

Lipenkov, V. Ya., A. A. Yekaikin, Yu. A. Shibayev, and I. N. Kuzmina (2003). Results of paleo-climatic studies from ice core of the deep borehole at Vostok station. *Express-Information, No. 16: Abstracts of the Concluding Session of the AARI Scientific Council*. St. Petersburg: AARI, pp. 84–85 [in Russian].

Loeng, H., V. Ozhigin, B. Ådlandsvik, and H. Sagen (1993). Current measurements in the northeastern Barents Sea. *ICES C. M.*, **1993-C**(41), 22.

Losev, S. M., Yu. A. Gorbumov, L. N. Dyment, and I. A. Sergeyeva (2005). Macro-deformations of the ice cover in the Arctic Basin. *Meteorology and Hydrology*, **8**, 38–50.

Luk'yanova, R. Yu. (2007). Modern studies on the problem of solar activity influence on climate variability. *AARI Proc.*, **447**, 210–226 [in Russian].

Mahoney, A. R., R. G. Barry, V. Smolyanitsky, and F. Fetterer (2008). Observed sea ice extent in the Russian Arctic, 1933–2006. *J. Geophys. Res.*, **113**(C11005), doi: 10.1029/2008JC004830.

Makarov, A. A. (1998). *World Energy and the Eurasian Energy Space*. Moscow: Energoatomizdat, 280 pp. [in Russian].

Makarov, V. I., and A. G. Tlatov (2000). Impact of solar activity on global warming of the oceans: The Sun in the maximum of activity and solar–terrestrial analogies. *Abstracts of GAO RAN, St. Petersburg, September 17–22*. St. Petersburg: Russian Academy of Science, p. 50 [in Russian].

Makshtas, A. P. (1984). *Heat Balance of the Arctic Ice in the Wintertime*. Leningrad: Gidrometeoizdat, 67 pp. [in Russian].

Makshtas, A. P., E. L. Andreas, and S. V. Shutilin (2001). Possible dynamic and thermal causes for the recent decrease in sea ice in the Arctic Basin. *Proceedings of the Sixth Conference on Polar Meteorology and Oceanography, San Diego, CA, May 14–18*, pp. 17–20.

Makshtas, A. P., S. V. Shutilin, and V. F. Romanov (2002). Sensitivity of modeled sea ice to external forcing and parameterizations of heat exchange processes. *Ice in the Environment: Proceedings of the 16th IAHR International Symposium on Ice, Dunedin, New Zealand, December 2–6*, pp. 90–98.

Makshtas, A., D. Atkinson, M. Kulakov, S. Shutilin, R. Krishfield, and A. Proshutinsky (2007). Atmospheric forcing validation for modeling the central Arctic. *Geophysical Research Letters*, **34**(L20706), doi: 10.1029/2007GL031378.

Maksimov, I. V. (1954). On some geophysical manifestations of the eleven-year cycle of solar activity. *Izvestiya AN SSSR Geography Series*, **1**, 15–32 [in Russian].

Maksimov, I. V. (1955). On some geophysical causes of multiyear fluctuations of sea-ice extent in the northern part of the Atlantic Ocean. *Scientific Notes of LVIMU*, **1**, 14–56 [in Russian].

Maksimov, I. V. (1960). Some issues of the study of multiyear fluctuations of the total sea-ice extent of the Arctic Seas. *Problems of the Arctic and the Antarctic*, **2**, 3–6 [in Russian].

Maksimov, I. V. (1970). *Geophysical Forces and Waters of the Ocean*. Leningrad: Gidrometeoizdat, 447 pp. [in Russian].

Maksimov, I. V., and V. P. Karklin (1969). Seasonal and multiyear changes in the geographical location and intensity of the Siberian maximum of atmospheric pressure. *Izvestiya VGO*, **101**(4), 320–330 [in Russian].

Maksimov, I. V., and V. P. Karklin (1970b). Seasonal and multiyear changes in the geographical location and intensity of the Azores maximum of atmospheric pressure. *Izvestiya AN SSS*, **1**, 17–23 [in Russian].

Maksimov, I. V., and V. P. Karklin (1970a). Seasonal and multiyear changes in the depth and the geographical location of the Aleutian Low of atmospheric pressure. *Izvestiya VGO*, **5**, 422–431 [in Russian].

Maksimov, I. V., and B. A. Sleptsov-Shevlevich (1963). On the study of eleven-year variations of atmospheric pressure in the Antarctic. *Information Bulletin of the Soviet Antarctic Expedition*, **43**, 5–10 [in Russian].

Maksimov, I. V., and B. A. Sleptsov-Shevlevich (1971). Solar activity and the baric field of the Earth. *Problems of the Arctic and the Antarctic*, **38**, 125–128 [in Russian].

Malmberg, S. (1969). Hydrographic changes in the waters between Iceland and Jan Mayen in the last decade. *Jöküll*, **19**, 30–43 [in Russian].

Manabe, S. and R. T. Wetherald (1975). The effect of doubling the CO_2 concentration on the climate of a general circulation model. *J. Atmos. Sci.*, **32**, 3–15.

Manabe, S., and R. T. Wetherald (1967). Thermal equilibrium of atmosphere with a given distribution of relative humidity. *J. Atmos. Sci.*, **24**, 241–259.

Mann, M. E., and P. D. Jones (2003). Global surface temperatures over the past two millennia. *Geophysical Research Letters*, **30**, 1820.

Mann, M. E., R. S. Bradley, and M. K. Hughes (1998). Global-scale temperature patterns and climate forcing over the past six centuries. *Nature*, **392**, 779–807.

Mann, M. E., R. S. Bradley, and M. K. Hughes (1999). Northern Hemisphere temperatures during the past millennium: Inferences, uncertainties, and limitations. *Geophysical Research Letters*, **26**, 759–762.

Mann, M. E., R. S. Bradley, and M. K. Hughes (2004). Corrigendum: Global-scale temperature patterns and climate forcing over the past six centuries. *Nature*, **430**, 105.

Mann, M. E., Z. Zhang, M. K. Hughes, R. S. Bradley, S. K. Miller, S. Rutherford, and F. Ni (2008). Proxy-based reconstructions of hemispheric and global surface temperature variations over the past two millennia. *Proceedings of the National Academy of Sciences of the U.S.A.*, **105**(36): 13252–13257, doi: 10.1073/pnas.0805721105.

McIntyre, S., and R. McKitrick (2003). Corrections to "Mann *et al.* (1998) Proxy data based and Northern Hemispheric average temperature series". *Energy and Environment*, **14**, 751–771.

McIntyre, S., and R. McKitrick (2005). *Hockey Sticks, Principal Components and Spurious Significance*, informal report. Available at *http://www.uoguelph.ca/rmckitri/research/trc.html*

McIntyre, S., and R. McKitrick (2006). *Surface Temperature Reconstructions for the Past 1,000–2,000 Years*, presentation. National Academy of Sciences Expert Panel, Washington, D.C. (March 2).

McIntyre, S., and R. McKitrick (2007). *The M&M Critique of the MBH98 Northern Hemisphere Climate Index: Update and Implications*, informal report. Available at *http://www.uoguelph.ca/rmckitri/research/trc.html*

McLaren, A. S., R. H. Bourke, J. E. Walsh, and R. L. Weaver (1994). Variability in sea ice thickness over the North Pole from 1958 to 1992. In O. M. Johannessen, R. D. Muench, and J. E Overland (Eds.), *Polar Oceans and Their Role in Shaping the Global Environment*. Washington, D.C.: American Geophysical Union, pp. 363–371.

Minobe, S. A. (1997). 50–70 year climatic oscillation over the North Pacific and North America. *Geophysical Research Letters*, **24**, 683–686.

Mironov, Ye. U. (2004). *Ice Conditions of the Greenland and Barents Seas and Their Long-range Forecast*. St. Petersburg: Gidrometeoizdat, 315 pp. [in Russian].

Mironov, Ye. U., and N. S. Uralov (1991). Year-to-year variations of ice transport from the Arctic Basin through the straits of the Canadian Arctic Archipelago and Fram Strait. *International Arctic Symposium*, IAHS No. 208. Oxford, U.K.: International Association of Hydrological Sciences, pp. 128–141.

Monin, A. S. (1969). *Weather Forecasting as a Goal of Physics*. Moscow: Nauka, 184 pp. [in Russian],

Monin, A. S. (2000). Influence of planets on the Earth's climate. *Global Environmental Changes (Climate and Water Regime)*. Moscow: Scientific World, pp. 122–128 [in Russian].

Monin, A. S., and Yu. A. Shishkov (1992). Dilemmas of warming in the 20th century. *Man and Elements*. Leningrad: Gidrometeoizdat, pp. 47–50 [in Russian].

Monin, A. S., and D. M. Sonechkin (2005). *Climate Fluctuations Based on Observations: Triple Solar and Other Cycles*. Moscow: P. P. Shirshov Institute of Oceanology/Russian Academy of Science/Nauka, 191 pp. [in Russian].

Moritz, R. E. (1988). *The Ice Budget of the Greenland Sea*, Technical Report APL-UWTR 8812. Seattle: Applied Physics Laboratory, University of Washington, 117 pp.

Mustel, E. P. (Ed.) (1974). *Solar–Atmospheric Relations in Climate Theory and Weather Forecasts*. Leningrad: Gidrometeoizdat, 483 pp. [in Russian].

Nagashima, T., H. Shiogama, T. Yokohata, S. A. Crooks, and T. Nozawa (2006). The effect of carbonaceous aerosols on surface temperature in the mid twentieth century. *Geophysical Research Letters*, **33**, L04702, doi: 10.1029/2005GL024887.

Nagovitsyn, Yu. A. (2007). Solar cycles during the Maunder Minimum. *Astronomy Letters*, **33**(5), 340–345.

Nagovitsyn, Yu. A., V. G. Ivanov, E. V. Miletsky, and D. M. Volobuev (2004). ESAI database and some properties of solar activity in the past. *Solar Physics*, **224**, 93–112.

Nansen, F. (1915). *To the Country of the Future*. Petrograd, 455 pp. [in Russian].

NCDC (National Climatic Data Center) (2007). *Climate of 2007: US and Global Climate Perspectives*. Available at *http://lwf.ncdc.noaa.gov/oa/climate/research/2007/perspectives. html*

NCDC (National Climatic Data Center) (2008). *Climate of 2008: US and Global Climate Perspectives*. Available at *http://lwf.ncdc.noaa.gov/oa/climate/research/2008/perspectives. html*

Nikiforov, Ye. G. (2006). *Sterodynamic System of the Arctic Ocean*. St. Petersburg: AARI, 176 pp. [in Russian],

Nikiforov, Ye. G., and A. O. Shpaikher (1980). *Patterns of Large-scale Fluctuations in the Hydrological Regime of the Arctic Ocean*. Leningrad: Gidrometeoizdat, 270 pp. [in Russian].

Nikolayev, Yu. V. (1963). On the theory of transformation of air masses over the sea. *Problems of the Arctic and the Antarctic*, **13**, 35–43 [in Russian].

Nikolayev, Yu. V. (1981). *Role of Large-scale Ocean–Atmosphere Interaction in the Formation of Weather Anomalies*. Leningrad: Gidrometeoizdat, 52 pp. [in Russian].

Nikolayeva, A. Ya., and N. P. Shesterikov (1970). Method for calculating ice conditions (using the example of the Laptev Sea). *AARI Proc.*, **292**, 143–217 [in Russian].

Nozawa, T., T. Nagashima, H. Shiogama, and S. A. Crooks (2005). Detecting natural influence on surface air temperature change in the early twentieth century. *Geophysical Research Letters*, **32**, L20719, doi: 10.1029/2005GL023540.

NSIDC (National Snow and Ice Data Center) (2007, 2008). *National Snow and Ice Data Center Notes*, **61**, 65, Fall. Available at *http://nsidc. org/pubs/notes/*

Ol', A. I. (1969). Indexes of the Earth's magnetic field perturbation and their solar–geophysical importance. *AARI Proc.*, **289**, 5–23 [in Russian].

Ol', A. I., and Sleptsov-Shevlevich B. (1972). Influence of a 22-year cycle of solar activity on the atmosphere of the northern hemisphere of the Earth. *Problems of the Arctic and the Antarctic*, **40**, 84–94 [in Russian].

Oreskes, N. (2004). The scientific consensus on climate change. *Science*, **306**, 1686.

Oreskes, N., E. M. Conway, and M. Shindell (2008). *From Chicken Little to Dr. Pangloss, William Nierenberg, Global Warming, and the Social Deconstruction of Scientific Knowledge*. Available at *http://www.lse.ac.uk/collections/CPNSS/projects/ContingencyDissent InScience/DP/DPOreskesetalChickenLittleOnlinev2.pdf*

Panov, V. V., and A. O. Shpaikher (1963). Influence of Atlantic water on some features of the hydrological regime of the Arctic Basin and the adjoining seas. *Oceanology*, **3**(4), 18–29.

Pogosyan, Kh. P. (1972). *General Circulation of the Atmosphere*. Leningrad: Gidrometeoizdat, 394 pp. [in Russian].

Pogosyan, Kh. P., and Z. L. Turketti (1970). *Atmosphere of the Earth*. Moscow: Prosveshcheniye, 320 pp. [in Russian].

Polyakov, I. V., and M. A. Johnson (2000). Arctic decadal and interdecadal variability. *Geophysical Research Letters*, **27**(24), 4097–4100.

Polyakov, I. V., A. Y. Proshutinsky, and M. A. Johnson (1999). Seasonal cycles in two regimes of Arctic climate. *J. Geophys. Res.*, **104**(C11), 25761–25788.

Polyakov, I. V., G. V. Alekseev, R. V. Bekryaev, U. S. Bhatt, R. L. Colony, M. A. Johnson, V. P. Karklin, A. P. Makshtas, D. Walsh, and A. V. Yulin (2002). Observationally based assessment of polar amplification of global warming. *Geophysical Research Letters*, **29**(18), 1878, doi: 10.1029/2001GL011111.

Polyakov, I. V., R. V. Bekryaev, G. V. Alekseev, U. S. Bhatt, R. L. Colony, M. A. Johnson, A. P. Makshtas, and D. Walsh (2003). Variability and trends of air temperature and pressure in the maritime Arctic, 1875–2000. *Journal of Climate*, **16**(12), 2067–2077.

Polyakov, I. V., G. V. Alekseev, L. A. Timokhov, U. S. Bhatt, R. L. Colony, H. L. Simmons, D. Walsh, J. E. Walsh, and V. F. Zakharov (2004). Variability of the intermediate Atlantic water of the Arctic Ocean over the last 100 years. *Journal of Climate*, **17**(23), 4485–4497.

Polyakov, I. V., A. Beszczynska, E. C. Carmack, I. A. Dmitrenko. E. Fahrbach, I. E. Frolov, R. Gerdes, E. Hansen, J. Holfort, V. V. Ivanov *et al.* (2005). One more step toward a warmer Arctic. *Geophysical Research Letters*, **32**(L17605), 1–4, doi: 10.1029/2005GL023740.

Ponomarev, V. I., V. V. Krokhin, D. D. Kaplunenko, and A. S. Salomatin (2003). Multiscale climate variability in the Asian Pacific. *Pacific Oceanography*, **1**(2), 125–137 [in Russian].

Ponomarev, V. I., D. D. Kaplunenko, and V. V. Krokhin (2005). Tendencies of climate changes in the second part of the 20th century in North East Asia, Alaska and the northwestern Pacific Ocean. *Meteorology and Hydrology*, **2**, 15–26 [in Russian].

Popov, F. V. (2000). Influence of flaw polynyas on weather formation and transformation of the thermal baric field of the Northern Polar Area. *ECIMO News*, **12**. Available at *http://www.oceaninfo.ru* [in Russian].

Porubayev, V. S. (2000). Influence of dynamic and thermal factors on seasonal changes in the average ice cover thickness in the Arctic Basin. *Meteorology and Hydrology*, **11**, 73–79 [in Russian].

Proshutinsky, A. Y. (1993). *Fluctuations of the Arctic Ocean Level*. St. Petersburg: Gidrometeoizdat, 216 pp. [in Russian].

Proshutinsky, A. Y., and M. A. Johnson (1997). Two circulation regimes of the wind-driven Arctic Ocean. *J. Geophys. Res.*, **102**, 12493–12514.

Proshutinsky, A. Y., I. V. Polyakov, and M. A. Johnson (1999). Climate states and variability of Arctic ice and water dynamics during 1946–1997. *Polar Research*, **18**(2), 135–142.

Proshutinsky, A. Y, R. H. Bourke, and F. A. McLaughlin (2002). The role of the Beaufort Gyre in Arctic climate variability: Seasonal to decadal climate scales. *Geophysical Research Letters*, **29**(23): 151–154, doi: 10.1029/2002GL015488.

Proshutinsky, A. Y., I. M. Ashik, E. N. Dvorkin, S. Häkkinen, R. A. Krishfield, and W. R. Peltier (2004). "100-Year" sea level change in the Russian sector of the Arctic Ocean. *J. Geophys. Res.*, **109**(C03042), doi: 10.1029/2003JC002007.

Quadfasel, D., A. Sy, D. Wells, and A. Tunik (1991). Warming in the Arctic. *Nature*, **350**(6317), 385, doi: 10.1038/350385a0.

Rakipova, L. R. (1962). Impact of climate change on Arctic ice. *Meteorology and Hydrology*, **9**, 28–30 [in Russian].

Rapp, D. (2008) *Assessing Climate Change*. Springer/Praxis, Heidelberg, Germany/Chichester, U.K.

Raspopov, O. M., V. A. Dergachev, and T. H. Kolstrom (2004). Cyclicity of polar activity and its relation to climate variability. *Solar Physics*, **224**, 455–463.

Reid, G. C. (2001). Solar variability and the Earth's climate: Introduction and overview. In E. Friis-Christensen, C. Fröhlich, J. D. Haigh, M. Schüssler, and R. von Steiger (Eds.), *Solar Variability and Climate*. Kluwer Academic/Space Science Series of ISSI, Vol. 11, pp. 1–11.

Rigor, I. G., J. M. Wallace, and R. L. Colony (2002). Response of sea ice to the Arctic Oscillation. *Journal of Climate*, **15**, 2648–2663.

Rothrock, D. A., Y. Yu, and G. A. Maykut (1999). Thinning of the Arctic sea-ice cover. *Geophysical Research Letters*, **26**(23): 3469–3472.

Rozhkov, V. A. (2001). *Theory and Methods of Statistical Assessment of Probabilistic Characteristics of Random Values and Functions with Hydrometeorological Examples*. St. Petersburg: Gidrometeoizdat, 340 pp. [in Russian].

Rudyaev, F. I., V. K. Trofimov., and Yu. A. Kravchuk (1985). Rhythmic changes of the Earth's rotation speed during the period 1664 to 1976. *Izvestiya VGO*, **117**(3): 252–257 [in Russian].

Ryabov, Yu. A. (1988). *Motions of Celestial Bodies*. Moscow: Nauka, 240 pp. [in Russian].

Saltzman, B., A. Sutera, and A. Evenson (1981). Structural stochastic stability of a simple auto-oscillatory climatic feedback system. *J. Atmos. Sci.*, **38**, 494–503.

Santsevich, T. I. (1970). On the methodology of long-range hydrometeorological forecasts for the Arctic. *AARI Proc.*, **292**, 49–86 [in Russian].

Santsevich, T. I., Z. M. Gudkovich, and V. P. Karklin (1979). On the relation of ice conditions of the Wrangel region to the indicators of atmospheric circulation of the preceding seasons. *AARI Proc.*, **363**, 30–39 [in Russian].

Schauer, U., B. Rudels, R. D. Muench, and L. Timokchov (1995). Circulation and water mass modification along the Nansen Basin slope. *Berichte zur Polarforschung*, **176**, 94–98.

Schulte, K. -M. (2008). Scientific consensus on climate change? *Energy and Environment*, **19**, 281–286.

Serreze, M. C., A. P. Barrett, A. G. Slater, R. A. Woodgate, K. Aagaard, R. B. Lammers, M. Steele, R. Moritz, M. Meredith, and C. M. Lee (2006). The large-scale freshwater cycle of the Arctic. *J. Geophys. Res.*, **111**(C11010), doi: 10.1029/2005JC003424.

Sherstyukov, B. G. (2008). *Regional and Seasonal Patterns of Modern Climate Change*. Obninsk: GU VNIGMI-MTsD, 247 pp. [in Russian].

Shimada, K., T. Kamoshida, M. Itoh, S. Nishino, E. Carmack, F. McLaughlin, S. Zimmermann, and A. Proshutinsky (2006). Pacific Ocean inflow: Influence on catastrophic reduction of sea ice cover in the Arctic Ocean. *Geophysical Research Letters*, **33**, L08605, doi: 10.1029/2005GL025624.

Shirochkov, A. V., and L. N. Makarova (1998). Long-term variability of the solar wind dynamic pressure and its climate consequences. Paper presented at *International Symposium on Space Plasma Studies by In-situ and Remote Measurements, Moscow, June 1–5* [in Russian].

Shoutilin, S. V., A. P. Makshtas, M. Ikeda, A. V. Marchenko, and R. V. Bekryaev (2005). Dynamic–Thermodynamic Sea Ice Model: Ridging and its application to climate study and navigation. *Journal of Climate*, **18**, 3840–3855.

Shuleikin, V. V. (1953). *Physics of the Sea*. Moscow: AN SSSR, 990 pp. [in Russian].

Shy, T. L., and J. E. Walsh (1996). North Pole ice thickness and association with ice ocean history 1977–1992, 1979–1986. *Geophysical Research Letters*, **23**, 2975–2978.

Sibirtsev, N., and V. Itin (1936). *The Northern Sea Route and the Kara Expeditions*. Novosibirsk: West Siberia Publishing House, 231 pp. [in Russian].

Singer, S. F. (Ed.) (2008). *Nature, Not Human Activity, Rules the Climate: Summary for Policymakers of the Report of the Nongovernmental International Panel on Climate Change*. Chicago: The Heartland Institute, 40 pp.

Sleptsov-Shevlevich, B. A. (1991). *Geophysical Bases of Marine Hydrological Forecasting*. Moscow: V/O Mortekhinformreklama, 103 pp. [in Russian].

Sleptsov-Shevlevich, B. A. (1996). Background forecast of ice area of the sub-Atlantic Arctic. *Izvestiya RGO*, **128**(2), 55–58 [in Russian].

Sleptsov-Shevlevich, B. A., and V. F. Zakharov (1996). Manifestation of solar activity in multiyear fluctuations of sea-ice extent in the sub-Atlantic Arctic. *Izvestiya RGO*, **128**(2), 55–58 [in Russian].

Smirnov, N. P., V. N. Vorobyev, and R. Yu. Kachanov (1998). *The North Atlantic Oscillation and Climate*. St. Petersburg: RGGTMU, 122 pp. [in Russian],

Smirnov, V. I. (1974). *Ice Conditions for Navigation of Ships in Waters of the Canadian–Alaskan Arctic*. Leningrad: Gidrometeoizdat, 180 pp. [in Russian].

Smolyanitsky, V. M. (2003). *Spatial–Temporal Variability of the Ice Cover Characteristics on the Basis of the "Global Digital Sea-Ice Data Bank"*, Abstract of a thesis for candidate of geographic science. St. Petersburg: AARI, 24 pp. [in Russian].

Solanki, S. K., I. G. Usoskin, B. Kromer, M. Schüssler, and J. Beer (2004). Unusual activity of the Sun during recent decades compared to the previous 11,000 years. *Nature*, **431**(7012), 1084–1087, doi: 10.1038/nature02995.

Sonechkin, D. M., N. M. Datsenko, and N. N. Ivaschenko (1997). Estimation of the global warming trend by wavelet analysis. *Izvestia, Atmospheric and Oceanic Physics*, **33**(2): 184–194.

Soon, W. (2005). Variable solar irradiance as a plausible agent for multidecadal variations in the Arctic-wide surface air temperature record of the past 130 years. *Geophysical Research Letters*, **32**(L16712), doi: 10.1029/2005GL023429.

Sorokhtin, O. G. (2001). Effect of greenhouse gases: Myth and reality. *Vestnik RAEN*, **1**(1), 8–21.

Stefan, J. (1891). Über die Theorie der Eisbilding, insbesondere über Eisbildung im Polarmeere. *Ann. Physik (3rd Ser.)*, **42**, 269–286.

Stuart, A., and J. K. Ord (1994). *Kendall's Advanced Theory of Statistics*, Vol. 1: *Distribution Theory*, Sixth Edition. London: Arnold, 676 pp.

Subbotin, V. V. (1988). On the issue of the role of sea ice in the dynamics of the climatic system of the Arctic. *AARI Proc.*, **404**, 82–96 [in Russian].

Swift, J. H., K. Aagaard, L. A. Timokhov, and E. G. Nikiforov (2005). Long-term variability of Arctic Ocean waters: Evidence from a reanalysis of the EWG data set. *J. Geophys. Res.*, **110**(C03012), 1–14.

Sytinsky, A. D. (1987). *Relation of the Earth's Seismicity to Solar Activity and Atmospheric Processes*. Leningrad: Gidrometeoizdat, 100 pp. [in Russian].

Teitelbaum, K. A. (1977). Dependence of air temperature over the Kara Sea in spring–summer on sea-ice extent and air transport. *AARI Proc.*, **346**, 109–117 [in Russian].

Teitelbaum, K. A. (1979). Dependence of air temperature over the Laptev, East-Siberian and Chukchi Seas in summer on sea-ice extent and air transport. *AARI Proc.*, **363**, 81–90 [in Russian].

Thompson, D. J. W., and J. M. Wallace (1998). The Arctic Oscillation signature in the wintertime geopotential height end temperature fields. *Geophysical Research Letters*, **25**, 1297–1300.

Timofeyev, V. T. (1960). *Water Masses of the Arctic Basin*. Leningrad: Gidrometeoizdat, 192 pp. [in Russian].

Torrence, C., and G. P. Compo (1998). A practical guide to wavelet analysis. *Bull. Amer. Meteor. Soc.*, **79**, 61–78.

Treshnikov, A. F., L. L. Balakshin, N. A. Belov, R. M. Demenitskaya, V. D. Dibner, A. M. Karasik, A. O. Shpaikher, and N. D. Shurgayeva (1967a). Geographical names of the main parts of Arctic Basin seabed relief. *Problems of the Arctic and the Antarctic*, **27**, 5–15.

Treshnikov, A. F., A. O. Shpaikher, and B. V. Gindysh (1967b). Heat exchange of the Southern Ocean with the atmosphere. *Problems of the Arctic and the Antarctic*, **27**, 35–47.

Usoskin, I. G., S. K. Solanki, M. Schussler, K. Mursula, and K. Alanko (2003). Millennium-scale sunspot number reconstruction: Evidence for an unusually active sun since the 1940s. *Physical Review Letters*, **91**(21), 211101–4.

Vangengeim, G. Ya. (1935). *Application of Synoptic Methods to Investigation and Characterization of Climate*. Leningrad: Gidrometeoizdat, 112 pp. [in Russian].

Vangengeim, T. G. (1986). Spectral analysis of the expansion coefficients of mean monthly anomalies of atmospheric pressure of the Northern Hemisphere by natural orthogonal components for the winter period. *AARI Proc.*, **393**, 131–137 [in Russian].

Vasilieva, G. Ya. (1997). Solar activity as manifestation of self-organization of the solar system. *Modern Problems of Solar Cyclicity*. St. Petersburg: Glavn. Astron. Obzerv., pp. 292–295 [in Russian].

Vasilieva, G. Ya., M. M. Nesterov, and Yu. V. Chernykh (2002). On the process of generation of the magnetic field on the Sun at the change of dynamic parameters of the Solar System. *Problems of the Study of the Universe*, **25**(II), 303–320 [in Russian].

Vinje, T. (1998). NAO winter index and the Nordic Sea Ice area in April. *ACSYS Arctic Forecast*, **3**, 4.

Vinje, T. (2000). Anomalies and trends of sea ice extent and atmospheric circulation in the Nordic Seas during the period 1864–1998. *Journal of Climate*, **2000**, 21.

Vinje, T. E., and O. Finnekasa (1986). Ice transport through Fram Strait. *Norsk Polarinstitute Skrifter*, **186**, 39.

Vinnikov, K. Y., A. Robock, R. J. Stouffer, J. E. Walsh, C. L. Parkinsson, D. J. Cavaliery, F. B. Mitchell, D. Garrett, and V. F. Zacharov (1999). Global warming and northern hemisphere sea ice extent. *Science*, **286**, 1934–1937.

Vitinsky, Yu. I. (1973). *Cyclicity and Forecasts of Solar Activity*. Leningrad: Nauka, 258 pp. [in Russian].

Vitinsky, Yu. I., A. I. Ol', and B. I. Sazonov (1976). *The Sun and the Earth's Atmosphere*. Leningrad: Gidrometeoizdat, 352 pp. [in Russian].

Vize, V. Yu. (1940). *Climate of the Soviet Arctic Seas*. Moscow: Glavsevmorput', 124 pp. [in Russian].

Vize, V. Yu. (1944a). Hydrological conditions in the ice edge area in the Arctic Seas. *AARI Proc.*, **184**, 125–151 [in Russian].

Vize, V. Yu. (1944b). Bases of long-range ice forecasts for the Arctic Seas. *AARI Proc.*, **190**, 274 [in Russian].

Vize, V. Yu. (1944c). *Fluctuations of Solar Activity and Sea Ice Extent of the Polar Seas*, presentation at the Jubilee Session of the Arctic Institute. Leningrad: Glavsevmorput', 7 pp. [in Russian].

Vize, V. Yu. (1948). *Soviet Arctic Seas*. Leningrad: Glavsevmorput', 414 pp. [in Russian].

Vize, V. Yu. (1951). Results of meteorological observations. *Proceedings of the Drifting Expedition of Glavsevmorput' onboard the Icebreaking Ship G. Sedov in 1937–1940*, Vol. 2, pp. 7–393.

Volkov, N. A. and Z. M. Gudkovich (1967). Main results of the studies of ice drift in the Arctic Basin. *Problems of the Arctic and the Antarctic*, **27**, 55–64 [in Russian].

Volkov, N. A., and B. A. Sleptsov-Shevlevich (1970). Two-year cycle in the fluctuations of sea-ice extent. *Problems of the Arctic and the Antarctic*, **34**, 13–19 [in Russian].

Volkov, N. A., and B. A. Sleptsov-Shevlevich (1971). On cyclicity in the sea-ice extent fluctuations in the Arctic Seas. *AARI Proc.*, **303**, 5–35 [in Russian].

Volkov, N. A., and V. F. Zakharov (1977). Ice cover evolution in the Arctic in connection with climate changes. *Meteorology and Hydrology*, **7**, 47–55 [in Russian].

Vorobiev, V. N., and N. P. Smirnov (2003). *The Arctic High and Climate Dynamics of the Northern Polar Area*. St. Petersburg: RGGMU, 82 pp. [in Russian].

Vowinchel, F. (1964). Ice transport in the East Greenland current and its causes. *Arctic*, **17**(2), 111–119.

Wadhams, P. (1990). Evidence for thinning of the Arctic ice cover north of Greenland. *Nature*, **345**, 795–797.

Wadhams, P. (1994). Variability in sea-ice thickness over the North Pole from 1958 to 1992. In O. M. Johannessen, R. D. Muench, and J. E. Overland Eds.), *The Polar Oceans and Their Role in Shaping the Global Environment*, Geophysical Monograph Series Vol. 85. Washington, D.C.: American Geophysical Union, pp. 337–361.

Wallace, J. M., Y. Zhang, and J. A. Renwick (1995). Dynamic contribution to hemispheric mean temperature trends. *Science*, **270**, 780–783.

Walsh, J. E., W. L. Chapman, and T. L. Shy (1995). Recent decrease in sea level pressure in the central Arctic. *Journal of Climate*, **9**(2), 480–486, doi: 10.1175/1520-0442(1996)009 < 0480: RDOSLP > 2. 0. CO;2.

Wegman, E. J., D. W. Scott, and Y. H. Said (2006). *Ad Hoc Committee Report on the Hockey Stick Global Climate Reconstruction*. Washington, D.C.: The Congressional Committee on Energy and Commerce (July 14). Available at *http://republicans.energycommerce.house. gov/108/home/07142006_Wegman_Report.pdf*

Westbrook, G. (1998). After Kyoto: Science still probes global warming causes. *Oil and Gas Journal*, **96**(3): 40–42.

Yegorov, E. G. (2004). Solar activity, baric waves in the surface atmosphere of the Arctic and multiyear changes in the Arctic Oscillation. *Meteorology and Hydrology*, **2**, 27–37 [in Russian].

Zakharov, V. F. (1976). Cooling in the Arctic and the ice cover of the Arctic Seas. *AARI Proc.*, **337**, 96 [in Russian].

Zakharov, V. F. (1977). Surface Arctic water as a factor of stability of the ice cover. *AARI Proc.*, **346**, 122–134 [in Russian].

Zakharov, V. F. (1978). *The World Ocean and the Pleistocene Glacial Epochs*. Leningrad: Gidrometeoizdat, 64 pp. [in Russian].

Zakharov, V. F. (1981). *Ice of the Arctic and Current Natural Processes*. Leningrad: Gidrometeoizdat, 136 pp. [in Russian].

Zakharov, V. F. (1996). *Sea Ice in the Climatic System*. St. Petersburg: Gidrometeoizdat, 213 pp. [in Russian].

Zakharov, V. F. (1997). *Sea Ice in the Climate System*, World Climate Research Programme/ Arctic Climate System Study, WMO/TD 782. Geneva: World Meteorological Organization, 80 pp.

Zakharov, V. F. and V. N. Malinin (2000). *Sea Ice and Climate*. St. Petersburg: Gidrometeoizdat, 92 pp. [in Russian].

Zavalishin, N. N., and G. M. Vinogradova (1990). On the relation of anomalies of monthly air temperatures to the Hale cycle and the dynamics of the distance between the Sun and the Earth. *SibNIGMI Proc.*, **93**, 25–32 [in Russian].

Zherebtsov, G. A., and V. A. Kovalenko (2001). Manifestation of solar activity in hydrometeorological characteristics of the Baikal region. *Studies on Geomagnetism, Aeronomy and Physics of the Sun*, **113**, 172–181.

Zubakin, G. K. (1987). *Large-scale Variability of the Ice Cover State in the Seas of the North European Basin*. Leningrad: Gidrometeoizdat, 160 pp. [in Russian].

Zubov, N. N. (1938). *Seawater and Ice*. Moscow: Glavsevmorput', 453 pp. [in Russian].

Zubov, N. N. (1944). *Ice of the Arctic*. Moscow: Glavsevmorput', 360 pp. [in Russian].

Index

(tables = *italics*, figures = **bold**)

Printing: Mercedes-Druck, Berlin
Binding: Stein+Lehmann, Berlin